M o n t e   Proceedings of the Centro Stefano Franscini
V e r i t à   Ascona

Edited by K. Osterwalder, ETH Zürich

Synchrotron Radiation: Selected Experiments in Condensed Matter Physics

Edited by
W. Czaja

1991

Birkhäuser Verlag
Basel · Boston · Berlin

Editor's address:

Prof. Dr. Wolfgang Czaja
Institut de Physique Appliquée
EPFL
1015 Lausanne
Switzerland

Deutsche Bibliothek Cataloging-in-Publication Data

Synchrotron radiation: selected experiments in condensed matter physics / ed. by W. Czaja. – Basel;
Boston; Berlin: Birkhäuser, 1991
(Monte Verità)

ISBN 978-3-7643-2594-7   ISBN 978-3-0348-7500-4 (eBook)
DOI 10.1007/978-3-0348-7500-4

NE: Czaja, Wolfgang [Hrsg.]

Contents

Preface

The present volume contains most of the invited talks and two contributed posters of a workshop which was held in July 1990 within the frame of the Centro Stefano Franscini, Monte Verità at Ascona in Switzerland.

The main idea has been to demonstrate that in many areas in solid state physics the use of synchrotron radiation as a powerful source for polarized uv radiation between 5 eV and about 1500 eV yields exciting new results. Furthermore, one has tried to attract a larger number of Ph D-students to bring them into contact with experienced people in the field.

This tutorial aspect, I suppose, is also in the spirit of Stefano Franscini, after whom the Centro, a joint undertaking between the Cantone Ticino and the Federal Institute of Technology in Zurich (ETHZ), has been named. Stefano Franscini, himself a dedicated teacher, has been the first member of his Canton in the Swiss Federal Council. As such he has been head of the departement of the interior and has been strongly supporting the foundation of the ETHZ. But he is also reknown, even today, as Cantonal Counceler, who has organised the elementary school system in Ticino.

Stefano Franscini 1796 - 1857

8

In organizing the scientific part of the workshop I had the benefit to draw on the longstanding experience of Y. Baer, Neuchâtel and A. Balzarotti, Roma. The «technical side» has been treated with general satisfaction by Katia Bastianelli, Zurich and Ascona, who has also helped in typing and retyping some of the manuscripts. I have to thank my collaborator, Cristina Marabelli-Bosio, who has been active in organising the workshop and also later on in preparing this volume. A financial contribution to the workshop by the Swiss Academy of Science in Bern is gratefully acknowledged. Last but not least I should like to thank Th. Hintermann from the Birkhäuser-Verlag, Basel, for his cooperation in completing this volume.

W. Czaja
Institut de Physique Appliquée, EPFL
CH-1015 Lausanne

Introduction

The field of synchrotron radiation is in an important stage of its historical development. Initiated in the late 1960's and early 1970's as a largely parasitic byproduct of elementary particle research, it soon demonstrated an almost virulent vitality of its own, leading to the implementation of a whole new generation of electron accelerators solely dedicated to synchrotron radiation activities. We are now at the threshold of a second revolution: several new sources are under development, and some of them push the performances to unprecedented levels, for example as far as brightness is concerned.

The intellectual and monetary investments in these new facilities have reached very high levels. The cost of the ELETTRA soft-x-ray source in Trieste will exceed 240 million dollars, and the hard-x-ray sources will reach the billion-dollar level. This spells the end of a myth: material science as "small" science, presumably as an alternate to "big" elementary-particle science -- and also poses a historical challenge to the synchrotron radiation community: using the new machines for what they are worth.

This book is, at least in part, an answer to the challenge. A group of prominent scientists from many countries were reunited for a few days in at atmosphere conducive to discussion, in the presence of students who are likely to become users of the new sources. They were able to present their ideas, to submit them to the scrutiny of their colleagues, and to extensive discussion. The chapters of this book are the distillate of this endeavor, covering diverse areas of synchrotron radiation research at the most advanced level.

The two most stimulating characteristic of the book are, in my opinion, the diversity of the areas covered by its different chapters, but also the specific emphasis in some fields that are the most likely to profit from the new sources under development

Experiments on magnetic properties, discussed in the first part of the book, are a good example of the latter. Recent results on the optical spectroscopy of magnetic systems with synchrotron radiation, obtained both in the USA and in Europe, have greatly augmented the general interest in this field. The lucid and extensive treatment provided by the chapter of Altarelli and Carra, and by that of Schütz, provide an excellent background for the understanding of these recent events and of their potential future developments. These are followed by more specialized presentation of G. Rossi and his coworkers; we call the reader's attention to their treatment of spin-polarization experiments, since this area, traditionally affected by low signal level, is one that stands to gain the most from the new sources; particularly important appear the present efforts to design and develop insertion-device-type sources of non-linearly-polarized synchrotron radiation.

The treatment of experiments on electronic structures is opened by the chapter on the band structure mapping of semimagnetic conductors, by Middelmann and Gumlich. The experiments discussed by the authors provide a nice example of the importance of synchrotron radiation in this kind of technique, when it is applied to all but the most elementary materials. Specifically, the tunability of the photon source is exploited for resonant photoemission experiments, that make the band.structure mapping data much more easy to interpret than those extracted for conventional photoemission. The intrinsic interest of the materials enhances the importance of the experiments presented by Middelmann and Gumlich.

The following chapter is on clusters experiments by authors of the Université de Lausanne and of the Kernforschungsanlage-Jülich. Once again the reader should pay full attention to the material that is presented: clusters are an extremely exciting area of research on, should we say, primordial solid-state systems. However, it has been strongly affected by the signal limitations of the present machines. The new sources will improve this situation by orders of magnitude, and the experiments described in this chapter are a good background for the understanding of the future opportunities.

Similarly stimulating is the chapter dedicated to liquid metals and alloys, by Indlekofer and Oelhafen. Because of its obvious experimental difficulties, this research is a newcomer in the field of synchrotron radiation: nothing was done before 1987, but the results obtained afterwards, including those presented here, demonstrate that these experiments have started at full steam. The following chapter, by de Groot, demonstrates how theory and experiment cross-fertilize each other in synchrotron radiation, and how such a cross-fertilization is enhanced by instrumentation advances. In this specific case, the instrumentation advances are related to the advent of the new monochromators of the class of the SX-700 and of the DRAGON, that make it possible to perform spectroscopy experiments with unprecedented levels of resolution; the chapter provides the theoretical background for some of these experiments.

The final chapter in the portion of the book dedicated to electronic structures discusses a specialized but very interesting application of the novel technique known as photoelectron diffraction, proposed by scientists from the University of Fribourg. Photoelectron diffraction was introduced a few years ago, by Dave Shirley, Neville Smith, Chuck Fadley, Dave Tong and other scientists; an alternate version known as "photoelectron EXAFS" had been proposed by Stoffel and Margaritondo in the late 1970's; more recently, there has been an evolution to photoelectron holography, through the work of Brian Tonner and other authors. All these varieties of photoelectron diffraction experiments have been negatively affected by low signal levels, and all are likely to profit from the new sources. The application presented in this chapter is an interesting new development that makes clever use of some of the potentialities of this approach.

Next is a chapter dedicated to interfaces. This is a field of synchrotron radiation research that is extensively covered by other reviews, and the present book wisely limits its coverage to some very recent studies. The author of the chapter, Cimino, is known (if not notorious) to his colleagues for having recently spread the panic by detecting previously neglected surface photovoltage effects; such effects have negatively affected some of the work done by other authors in recent years, and Cimino's detection (in cooperation with K. Horn and other scientists) has gained him the gratitude of science but not of all of his colleagues. Reading the chapter written by the co-author of this time-bomb is certainly a very stimulating experience.

Like in the case of interfaces, crystal structure studies are well covered by many other reviews, and therefore limited to one chapter here, by Weber of the Université de Lausanne. The solution, however, is different from that adopted in the case of interfaces: Weber's chapter is a rather extensive and exhaustive treatment of this field. It lucidly presents the advantages brought by the advent of synchrotron radiation in crystallography, with general considerations and specific examples. I personally found reading Weber's treatment a real intellectual pleasure.

In the development of a book like the present one, deciding what is excluded is almost as important as deciding what is included. The book, for example, is not an elementary introduction to synchrotron radiation. There are other books that fulfill that role, including one that modesty prevents me from citing explicitly. The present book builds in the direction of the future, assuming a general background knowledge of synchrotron radiation by the reader. Whereas it does not begin from the beginning, it does, instead, end at the end. This being at the present time the impending advent of the new, ultrabright sources.

The final chapter of the book is indeed specifically dedicated to the opportunities created by the new facilities. The author, Renzo Rosei, is highly qualified for treating this topic, since he is in charge of scientific activities at the Sincrotrone Trieste, the private company responsible for the commissioning of Trieste's source ELETTRA. The chapter reflects one of the most fascinating aspects of the author's personality: his capability to expand the scientific interests well beyond his specific and specialized domain (solid-state physics), and into diverse areas such as the life sciences and medicine. As a footnote to this chapter, I would like to mention the fact that Rosei's talk during the conference was delivered in an unprecedented mixture of English and Italian (the latter being required for the "external" audience to whom the presentation was primarily dedicated). The mixture did not touch the clarity of the presentation, made more interesting by the display of colorful and spectacular pictures produced by a variety of recent synchrotron radiation experiments.

As a sidenote to this final chapter, I cannot resist the temptation to add a few things of my own, again to demonstrate how rapid is the instrumentation progress in synchrotron radiation -- and also, again, as background to the new opportunities opened by the sources under development. Figure 1 illustrates the stunning changes made possible in synchrotron-radiation photoemission by instrumentation refinements of the past two years. primarily stimulated by the needs of experiments on high-temperature superconductors.

Both Fig. 1a and 1b show spectra taken at temperatures above and below the superconducting transition on BCSCO, in the energy region near the superconductivity gap. Notice the spectral changes, and in particular the shift in the leading edge that reflects the opening of the superconductivity gap. The differences between the two figures are in resolution: Fig 1a is from two years ago, taken with no angular resolution and with the typical hundred-millivolt-level resolution of most valence-band photoelectron experiments. Figure 1b was taken with high angular resolution, and with an energy resolution of 17 meV.

These improvements are very painful to obtain because of the low signal level, and of the consequently very long time required for each spectrum. New sources like ELETTRA and the ALS-Berkeley will produce undulator radiation with a brightness increase of 4-5 orders of magnitude with respect to the situation of Fig. 1: the reader can imagine the impact of this step change. I would also like to mention that the results of Fig. 1 are at home in a book on a conference in Switzerland, since the push for high resolution was led by Neuchatel scientists and notably by Yves Baer.

Figure 2 illustrates another novelty in synchrotron radiation photoemission: the advent of lateral resolution. This is a welcome improvement in a field that ordinarily integrates over areas of the order of millimeters. In Fig. 2 we see instead a total-yield photoelectron *micr*ograph, illustrating the fine details of a neuron network prepared by G, De Stasio of Frascati; the image was taken with 0.5-micron lateral resolution, using the spectromicroscope MAXIMUM developed at the Wisconsin Synchrotron Radiation Center by a cooperation team involving Berkeley, Wisconsin, Stanford, Minnesota, Xerox and the Ecole Polytechnique Fédérale of Lausanne. The time per micrograph is several hours, due again to low signal level (in spite of the use of an undulator); with ELETTRA, ALS and other sources, it could become much shorter, even with higher lateral resolution.

Facts like those discussed by Rosei or illustrated by Figs. 1 and 2 are the best source of optimism for the future of this field. The present book is a good, advanced guide towards this ultrabright future (the pun *is* intended). Before releasing the readers to some stimulating reading, I would like to conclude this introduction by expressing my personal gratitude to W. Czaja for having gone through the pain and agony of organizing the conference and developing the book, essentially single-handed. I am sure that this is not merely my personal feeling, and that the entire scientific community join me in telling him: thanks for a well-done job that was really needed.

Giorgio Margaritondo
Institut de Physique Appliquée
Ecole Polytechnique Fédérale
CH-1015 Lausanne, Switzerland

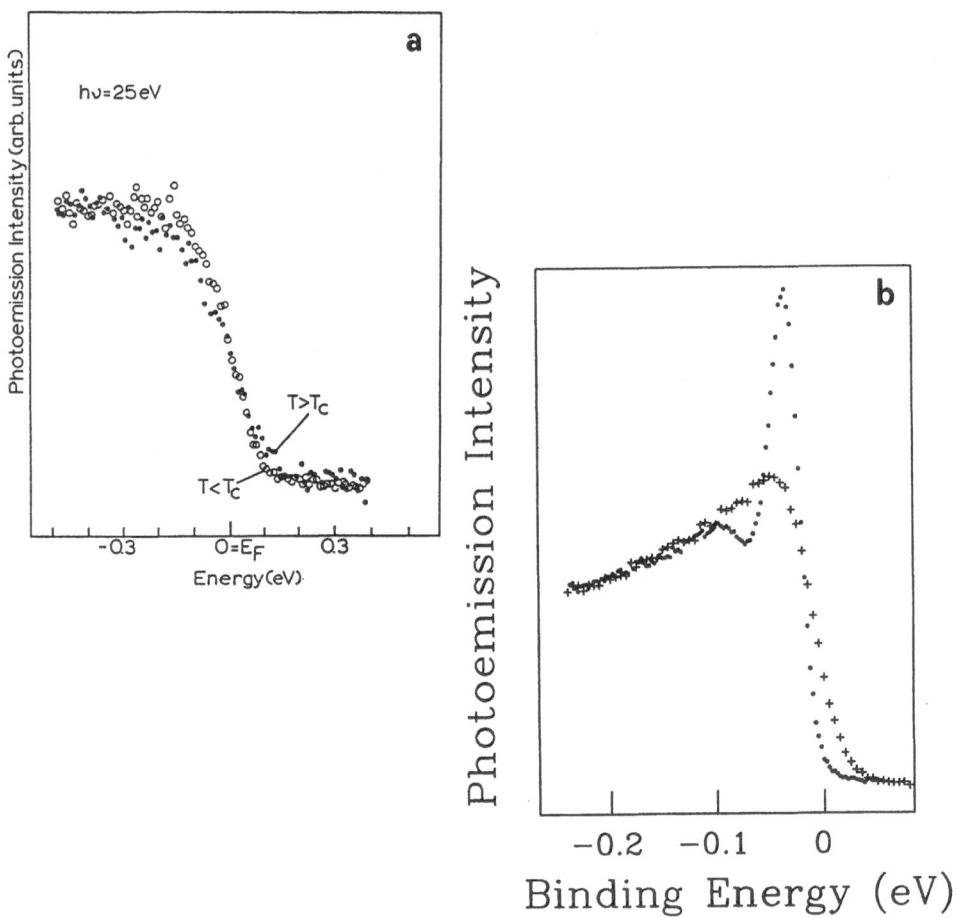

Fig. 1 - (a) Photoelectron spectra in the spectral region near the Fermi energy, taken for the high-temperature superconductor BCSCO at temperatures above (full circles) and below (open circles) the critical value (data from: Y. Chang, Ming Tang, R. Zanoni, M. Onellion, Robert Joynt, D. L. Huber, G. Margaritondo, P. A. Morris, W. A. Bonner, J. M. Tarascon and N. G. Stoffel, Phys. Rev. **B39**, 4740 (1989); (b) similar spectra, again above (crosses) and below (full circles) the critical temperature, taken more recently with much higher energy resolution, and higher angular resolution (Y. Hwu, L. Lozzi, M. Marsi, S. La Rosa, M. Winokur, P. Davis, M. Onellion, H. Berger, F. Gozzo, F. Lévy and G. Margaritondo, unpublished results).

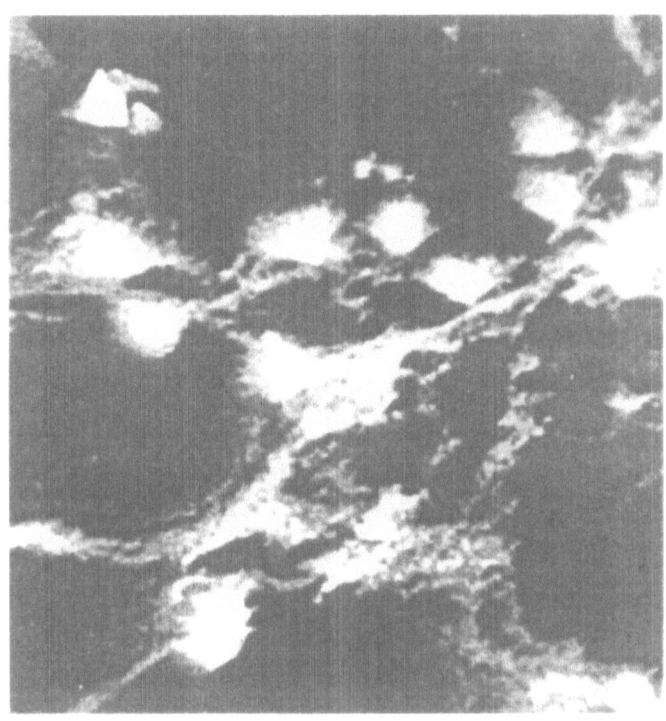

Fig. 2 - Photoelectron micrograph of a neuron network, taken with a lateral resolution of 0.5 micron.(data from: Delio Mercanti, Gelsomina De Stasio, M. Teresa Ciotti, C. Capasso, W. Ng, A. K. Ray-Chaudhuri, S. H. Liang, R. K. Cole, Z. Y. Guo, J. Wallace, G. Margaritondo, F. Cerrina, J. Underwood, R. Perera and J. Kortright, J. Vac. Sci. Technol. (in press)).

Magnetic Properties

DICHROISM EFFECTS IN THE X-RAY SPECTROSCOPY OF MAGNETICALLY ORDERED SYSTEMS

Massimo Altarelli* and Paolo Carra

ESRF, B.P. 220, F-38043 Grenoble, France

*   also at Max-Planck-Institut für Festkörperforschung, Hochfeld-Magnetlabor, B.P. 166X, F-38042 Grenoble, France

SUMMARY:

Theoretical results on linear X-ray dichroism in antiferro- or ferromagnetic systems and circular dichroism in ferromagnets are briefly described, with the aim of providing an interpretation of recent experiments.

INTRODUCTION:

Until recently, X-rays were hardly considered a valuable tool for the investigation of magnetic order. The experiments pioneered by de Bergevin and Brunel [1] have shown, since 1972, that photons couple to electron spins, as indeed predicted from quantum electrodynamics, and that, as a consequence, X-ray diffraction can reveal the periodicity of magnetic structure, e.g. in an antiferromagnet. However, these relativistic effects are small, such that their observation with laboratory sources was a real "tour de force". The advent of synchrotron radiation sources allows an easier observation of such effects [2], but magnetic diffraction of X-rays remains a technique unfrequently used, if compared with neutron diffraction. Very recently, from 1986 on, a remarkable development in the application of synchrotron radiation to magnetic systems has taken place. This is to be ascribed to two facts:

a)   The exploitation of the polarization properties (linear in the orbital plane, circular above or below such plane) of synchrotron radiation. A source of polarized X-rays makes magneto-optical techniques available, such as:
   - Magnetic Dichroism (linear or circular), Magneto-optic Kerr effect
   - Faraday effect.

These effects are not due to small relativistic terms, but to electric multipolar terms, of much greater intensity.

b)   The discovery in 1988 by Gibbs et al. [3, 4] of Resonant Exchange Scattering of X-rays, a process which yields magnetic information, with a cross section at least 3 orders of magnitude bigger than conventional magnetic diffraction. In this case, one takes advantage not only of the polarization features, but also of the "white" spectrum of synchrotron radiation, which allows selection of photons in resonance with an absorption threshold of a magnetic ion.

In addition techniques such as spin-dependent Compton scattering and spin-resolved photoemission are also becoming more common.

In the present paper, we shall discuss the salient features of magnetic dichroism experiments and of their theoretical interpretation.

MAGNETIC X-RAY DICHROISM:

Let us consider the X-ray photon absorption processes by an electronic system. In the standard weakly relativistic description of the interaction between radiation and matter (see e.g. [5]) they are controlled by the Hamiltonian:

$$H = -\frac{e}{mc} \sum_j \vec{A}(\vec{r}_j) \cdot \vec{p}_j \tag{1}$$

where the sum is over all electrons, $\vec{A}$ is the vector potential of the radiation field in the transverse gauge and $\vec{r}_j$, $\vec{p}_j$ are position and momentum of the j-th electron. The absorption cross section is easily expressed by inserting Eq. (1) into Fermi's golden rule.

Elementary considerations suggest that in a magnetic system absorption probabilities depend on the photon polarization. The Zeeman effect, which consists in the splitting of absorption lines for different polarizations, in the presence of an external magnetic field, has indeed a counterpart for the case of the internal field in a ferromagnet: it is the magneto-optic Kerr effect. The same holds for the Faraday effect, because the polarization dependence of the index of refraction corresponds, via the Kramers-Kronig relations, to a polarization dependence of the absorption coefficient.

Magnetic X-ray dichroism was predicted in 1985 by Thole et al. [6], on the basis of an atomic model for the 4f electrons in the rare earths. Exploiting the linear polarization of synchrotron radiation in the orbital plane, the effect was indeed observed, by the same group [7], in 1986 at the M_4, M_5 threshold of Tb in $Tb_3Fe_5O_{12}$, at an energy around 1.2 keV.

The circular X-ray dichroism was observed for the first time in 1987, exploiting the circular polarization above or below the orbital plane, by Schütz et al. [8] at the Fe K edge and was later measured by the same and by other groups in a variety of ferro- and ferrimagnetic systems.

In order to get some insight on the relationship between these experiments we expand the vector potential $\vec{A}$ of Eq. (1) in its electric and magnetic multipole components, i.e. we expand the photon 'wavefunction' in eigenfunctions of the angular momentum $\vec{L}$ and of its projection M with parity $(-1)^L$ (electric multipole) or $(-1)^{L+1}$ (magnetic multipole) [9]. For photons of frequency ω, with polarization $\hat{e}$, the electric dipole absorption is proportional to [10]:

$$W_{E1} = \frac{3c}{4\omega} \left\{ \left(w_{1,1} + w_{1,-1}\right) - i\left(\hat{e}^* \times \hat{e}\right) \cdot \hat{z}\left(w_{1,1} - w_{1,-1}\right) \right.$$

$$\left. + \left|\hat{e} \cdot \hat{z}\right|^2 \left(2 w_{1,0} - w_{1,1} - w_{1,-1}\right) \right\} \tag{2}$$

where w_{LM} is the matrix element, between initial and final states that conserve energy, of the electric 2^L-pole operator

$$Q_{LM} = -\frac{e}{(2L+1)!!} \sum_j r_j^L Y_{LM}\left(\hat{r}_j\right) \tag{3}$$

In Eq. (2), the unit vector $\hat{z}$ denotes the quantization axis which, for a magnetic ion, is conveniently chosen along the magnetization. If there is no magnetic order, the average (over all ions) of the second term is zero whatever the polarization $\hat{e}$, while the average of the third term is $(1/2)$ $(2 w_{1,0} - w_{1,1} - w_{1,-1})$, thus suppressing any $\hat{e}$ dependence of the absorption. On the other hand, for a ferromagnetic system (magnetization along $\hat{z}$ of all ions) or an antiferromagnetic one (magnetization along $\pm \hat{z}$) let us consider linear polarization, parallel or perpendicular to $\hat{z}$. In both cases $\hat{e}$ is real, so that $\hat{e}^* \times \hat{e} = 0$. We find:

$$W_{E1}\left(\hat{e} \,/\!/\, \hat{z}\right) = \frac{3c}{2\omega} w_{10}$$

$$W_{E1}\left(\hat{e} \perp \hat{z}\right) = \frac{3c}{4\omega}\left(w_{1,1} + w_{1,-1}\right) \tag{4}$$

In the case of left or right circular polarization:

$$\hat{e}^+ = \frac{-i}{\sqrt{2}}\left(\hat{e}_1 + i\,\hat{e}_2\right), \hat{e}^- = \frac{i}{\sqrt{2}}\left(\hat{e}_1 - i\,\hat{e}_2\right),$$ the circular dichroism is:

$$W_{E1}\left(\hat{e}^{+}\right) - W_{E1}\left(\hat{e}^{-}\right) = \frac{3\,c}{2\,\omega}\left(\hat{e}_1 \times \hat{e}_2\right) \cdot \hat{z}\left(w_{1,1} - w_{1,-1}\right) \tag{5}$$

In Eq. (5), $\hat{e}_1$ and $\hat{e}_2$ are two orthogonal unit vectors in the plane perpendicular to the photon wavevector. It is clear that circular dichroism can only occur in ferromagnetic (and ferrimagnetic) systems.

What is the physical information contained in a dichroism measurement? To clarify this point, it is useful to consider for a while the simplest case, in which each magnetic ion is regarded as a system of spherical symmetry, i.e. with complete neglect of crystal field effects. This is the "atomic" approximation (very good for the f levels of rare earths, but much less for the d levels of transition metals) adopted in Ref. [6]. In such situation it is easy to see, applying the Wigner-Eckart theorem, that:

$$w_{1,1} - w_{1,-1} \approx <r> \frac{M_0}{\left(J_0 + 1\right)\left(2\,J_0 + 1\right)} \tag{6}$$

for an ion in an initial state J_0, M_0, performing a transition to a state with $J_f = J_0 + 1$. Therefore, circular dichroism is proportional to the radial matrix element $<r>$ and to the component M_0 of J_0 along the magnetization axis. The same holds for $J_0 \rightarrow J_0 - 1$ and $J_0 \rightarrow J_0$ transitions. For linear dichroism, one finds instead:

$$2\,w_{10} - w_{1,1} - w_{1,-1} \approx <r> \left(M_0^2 - \frac{1}{3}J_0\left(J_0 + 1\right)\right) \tag{7}$$

Linear dichroism is therefore proportional to the deviation, with respect to the statistical average $1/3\,J_0\,(J_0 + 1)$, of the square of the J_0 component along the magnetization direction, and can be observed in antiferromagnets as well as ferromagnets. In Ref. [10], relations analogous to Eq.s (4) and (5) are deduced for quadrupole transitions.

To interpret the experiments performed so far on a variety of magnetic systems (and that display dichroism effects up to [11] 20% difference between circular polarizations at the L_2, L_3 edges of the 5d metal in OsFe, PtFe and IrFe alloys) the atomic approach is not always viable. Ebert et al. [12] adopted the computational techniques of electronic structure calculations based on the Local Density Approximation, and discussed the picture in which circular dichroism results from selection rules which, when applied to spin-orbit split core levels, favour final states with spin up for one polarization, and with spin down for the other. The different density of available final states, at a given energy, for the two spin directions in ferromagnets accounts for the difference in absorption.

An interplay of atomic-like and band effects was recently suggested to interpret [13] the circular dichroism at the $L_{2,3}$ edges of the ferromagnetic rare earths Gd and Tb [14]. Pre-edge structure in the circular dichroism spectra at $L_{2,3}$ edges is ascribed to 2p $\rightarrow$ 4f electric quadrupole transitions, whereas above the edge dipole transitions to the broad 5d band dominate. In this picture, the energy of the 4f levels is shifted downwards by several eV's by the interaction with the core hole. This is due to their strongly localized wavefunctions, such that the quadrupole 2p $\rightarrow$ 4f spectrum is best described by atomic-like calculations. Band theory, on the other hand, accounts for the higher energy dipole transitions to the 5d bands with remarkable accuracy. This combined atomic + band approach gives good agreement with experiment. It also leads to specific predictions concerning the dependence of the spectrum on the angle between the photon wavevector and the magnetization direction, which have not yet been confirmed experimentally.

References

[1]   F. de Bergevin and M. Brunel, Phys. Letters A39, 141 (1972)

[2]   D. Gibbs, D.E. Moncton, K.L. D'Amico, J. Bohr and B. Grier, Phys. Rev. Lett. 55, 234 (1985)

[3]   D. Gibbs, D.R. Harshman, E.D. Isaacs, D.B. Mc Whan, D. Mills and C. Vettier, Phys. Rev. Lett. 61, 1241 (1988)

[4]   J.P. Hannon, G.T. Tramnel, M. Blume and D. Gibbs, Phys. Rev. Lett. 61, 1245 (1988); (E) Phys. Rev. Lett. 62, 2644 (1989)

[5]   M. Blume, J. Appl. Phys. 57, 3615 (1985)

[6]   B.T. Thole, G. Van der Laan and G.A. Sawatzky, Phys. Rev. Lett. 55, 2086 (1985)

[7]   G. Van der Laan, B.T. Thole, G.A. Sawatzky, J.B. Goedkoop, J.C. Fuggle, J.–M. Esteva, R. Karnatak, J.P. Remeika and H.A. Dabkowska, Phys. Rev. B34, 103 (1986)

[8]   G. Schütz, W. Wagner, W. Wilhelm, P. Kienle, R. Zeller, R. Frahm and G. Materlik, Phys. Rev. Lett. 58, 737 (1987)

[9]   See e.g. V.B. Berestetskii, E.M. Lifshitz and L.P. Pitaevskii, Relativistic Quantum Theory, (Pergamon, New York, 1971)

[10]   P. Carra and M. Altarelli, Phys. Rev. Lett. 64, 1286 (1990)

[11]   G. Schütz, R. Wienke, W. Wilhelm, W. Wagner, P. Kienle, R. Zeller and R. Frahm, Z. Phys. B75, 495 (1989)

[12]   H. Ebert, P. Strange and B.L. Gyorffy, J. Appl. Phys. 63, 3055 (1988); Z. Phys. B73, 77 (1988)

[13]   P. Carra, B.N. Harmon, B.T. Thole, M. Altarelli and G.A. Sawatzky, to be published

[14]   G. Schütz, M. Knülle, R. Wienke, W. Wilhelm, W. Wagner, P. Kienle, R. Frahm, Z. Phys. B73, 67 (1988)

MAGNETIC PHOTOABSORPTION
WITH CIRCULARLY POLARIZED X-RAYS

Gisela Schütz
Fakultät für Physik der Technischen Universität München,
James-Franck-Strasse, D-8046 Garching, Germany

Abstract:

Spin-dependent X-ray absorption, which is also called circular magnetic X-ray dichroism (CMXD), is a new method to study ferromagnetic materials with circularly polarized synchrotron radiation in the X-ray energy range. The CMXD can show large relative magnetic effects amounting to 20 % and is an universal phenomenon found at inner-shell absorption edges in nearly all ferro(i)magnetically oriented atoms in magnetic media. The CMXD, which is the absorption counterpart to the magnetic resonance scattering, can in case of itinerant ferromagnets be well understood in the band-structure picture reflecting directly the spin-polarization of the final states populated in the absorption process.

1 Introduction

Since a long time magnetic photoabsorption or circular magnetic dichroism is a well known phenomenon of the interaction of photons with magnetic materials for visible light known as magneto-optical Kerr effect. Erskine and Stern have predicted [Ers75], that these phenomena should be also observable for higher photon energies ($E\gamma \sim 60..70$ eV) at the $M_{2,3}$-edges of transition metals and are related to the unfilled d-band spin polarization. Following the calculations of Thole et al. [Tho85] strong magnetic X-ray dichroism effects should also occur at the $M_{4,5}$-edges of magnetic rare-earth materials ($E\gamma \sim 800 - 1600$ eV), which was experientially confirmed for linearly polarized light [Laa86].

The existence of **Circular Magnetic X-ray Dichroism (CMXD)** or spin-dependent absorption, measured as difference of the absorption of right and left circularly polarized hard X-rays ($E\gamma > 5$ keV) at inner-shell absorption edges in ferromagnets, has been experimentally established for the first time at the K-edges of iron [Sch87]. It was also found at the L-edges of Gd and Tb metal [Sch88] and 5d-impurities in ferromagnetic hosts [Sch89a, Sch90a]. In the meanwhile it has been verified for about more than 100 studied systems at K-edges [Col89, Sch89b-e, Cla90, Bau90a-b] and L-edges of 3d- [Che90, Set90], 5p- [Ebe91b], 4f- [Sch88, Sch89b-d, Fis90, Bau90a-b] and 5d-elements [Sch89a-e, Ebe90a, Sch90a-c] and M-edges of 4f-elements [Set90], that CMXD is an universal phenomenon occuring at the inner-shell absorption edges of atoms oriented in ferro(i)magnetic systems, as pure metals, oxids, compounds and multilayered structures.

In this review the experimental methods to study CMXD are briefly described. Typical spin-dependent spectra in the hard X-ray range $E\gamma \gtrsim 5$ keV are presented and compared with the results of newly developed fully relativistic theories. In the framework of simplified pictures the physical origin of the CMXD is explained. The phenomenon of CMXD at inner-shell absorption-edges, which is strongly correlated to its diffractive counterpart the enhanced magnetic resonance scattering of linearly polarized X-rays [Han88, Car89], is discussed in view of its content on information on local microscopic magnetic structures.

2 Experimental aspects

The CMXD is measured by detecting the difference of the absorption rate or the absorption coefficient μ of circularly polarized photons for parallel (μ^+) and antiparallel (μ^-) orientation of the photon spin and (normally) the spin of the magnetic or majority like electrons in the ferro- or ferrimagnetic absorbers. Up to now almost all methods use circularly polarized synchrotron radiation emitted at small vertical angles with respect to the plane of the electron orbit. Since 1989 novel insertion devices like the helical wiggler at the Photon factory [Yam89] and the asymmetric wiggler at HASYLAB [Pfl90a, b] have been successfully tested as new sources with a sufficiently high circularly polarized flux. The experimental spectroscopic techniques correspond to those applied in the conventional absorption spectroscopy. For photon energies $E\gamma \gtrsim 5$ keV at HASYLAB a fast and effective method has been developed measuring simultaneously the transmission of two beams with opposite sense of circular polarization emitted above and below the synchrotron plane [Sch89e]. Small differences of the CMXD in one atom in different chemical compounds can be studied using target and a well known reference sample in a row. At LURE the energy-dispersive method, suitable for small sample sizes, has been also successfully applied for transmission measurements [Dar86, Bau90a, b]. For thicker multicomponent samples, especially single crystals or permanent magnetic alloys as well as thin layers on substrates, the detection of the characteristic fluorescence light with high signal to noise ratio can be a suitable, effective method [Col89]. At the NSLS the CMXD studies have been successfully extended towards the soft X-ray energy range 700 eV - 1.5 keV by monitoring the absoption rate via the detection of the photoelectron current from single crystal samples at normal incidence [Che90, Set90].

The difference of the absorption $\mu^+(E) - \mu^-(E)$ can be detected either by changing the sense of the photon circular polarization for fixed magnetization relative to the photon k-vector or by reversing the component of the sample magnetization in the photon beam direction for fixed sense of the photon polarization. In principle up to now the latter method has been used, which is much easier to realize technically. To deduce the systematics of the CMXD it is of advantage to introduce the «spin-dependent absorption profile», defined as the thickness independent value $\mu_c/\mu_0 = (\mu^+ - \mu^-)/(\mu^+ + \mu^-)$, which corresponds to the relative magnetic absorption cross-section. To allow a quantitative interpretation of the data especially in comparison with theory the spectra are normalized for complete degree of circular polarization ($P_c = 0$) and complete saturation magnetization. Due to uncertainties of these rescaling factors the data may contain errors in the vertical scale amounting to about 20 % in some cases. The first inflection point of the normal absorption profile is chosen as the origin of the energy scale.

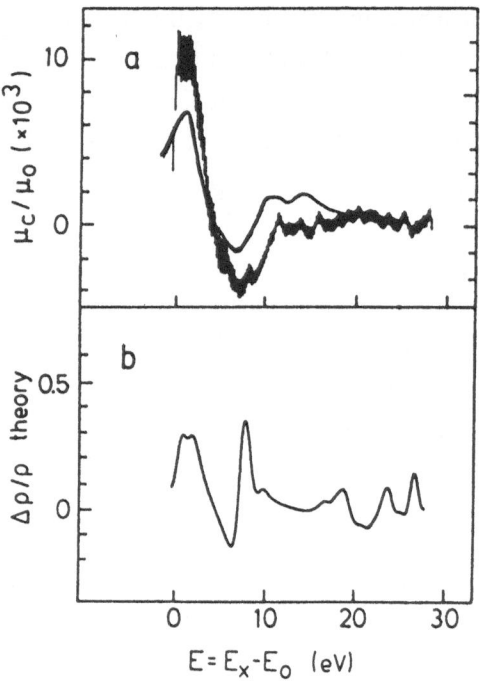

Fig.1 : a) Theoretical (solid line, from [Ebe88a,b]) and experimental spin-dependent absorption profile at the K-edge of Fe metal
b) Calculated spin-density profile $\Delta g/g$ of the unoccupied p-projected states at the Fermi level in ferromagnetic iron metal

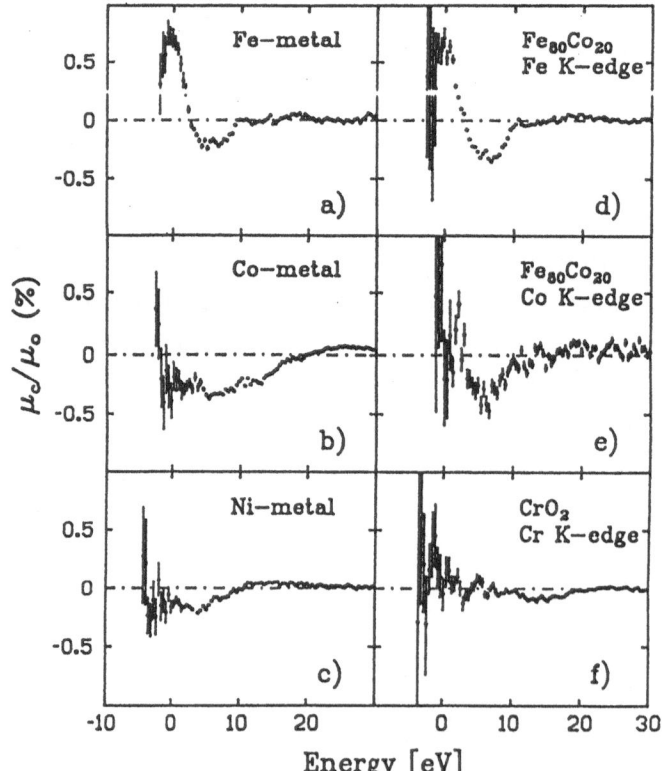

Fig. 2: Spin-dependent absorption profiles at the K-edges of metallic Fe (a), Co (b), Ni (c), in the alloy $Fe_{80}Co_{20}$ at the Fe K-edge (d), Co K-edge (e) and at the Cr K-edge (f) in the ferromagnetic oxide CrO_2.

3 Results and discussion

3.1 K-edges of 3d elements

For the first time the existence of CMXD has been confirmed at the K-edges of ferromagnetic iron measured at HASYLAB using the two-beam inclined-view method. The results have prompted first fully relativistic calculations of the absorption of circularly polarized X-ray in ferromagnetic media [Ebe88a, b]. In Fig. 1a the experimental and theoretical Fe K-CMXD spectra $[\mu_c/\mu_0](E)$ are presented. The theoretical spectra are convoluted with a Lorentzian (FWHM~1eV) and a Gaussian (FWHM ~ 1 eV) to take into account the core-hole life-time broadening and the experimental resolution. Apart from some differences at higher energies (E~8 eV) a qualitative good agreement between the experimental findings and theoretical predictions is found.

Also another important correlation could be deduced from this first experiment: In Fig. 1b the normalized difference of the density for the empty spin-up (ρ^+) and spin-down (ρ^-) p-projected final states $(\Delta\rho/\rho) = (\rho^+ - \rho^-)/(\rho^+ + \rho^-)$ deduced from results of spin-polarized band-structure calculations for iron metal is shown. The similarity of the experimental as well as the theoretical $[\mu_c/\mu_0](E)$-spectra and the «spin density»-profile $[\Delta\rho/\rho](E)$ show, that the CMXD signal is obviously correlated with the spin structure of the unoccupied states at the Fermi level. This important aspect will be discussed in the following chapter.

In all ferro(i)magnetic systems the states at the Fermi level E_F will be spin-split, which will be correlated with a spin density of the unoccupied states close to E_F characteristic for the magnetic system. Therefore, CMXD should occur in all ferro(i)magnetically oriented atoms. Despite the smallness of the magnetic effects, which are in the order of $\mu_c/\mu_0 \lesssim 1$ % at K-edges, this could be easily experimentally verified using the two-beam transmission method [Sch89e]. Some typical CMXD spectra measured at the K-edge in elementary Fe, Co (fcc), Ni, CoFe alloy and CrO_2, taken during a measuring time of about 2 hours, are shown in Fig. 2.

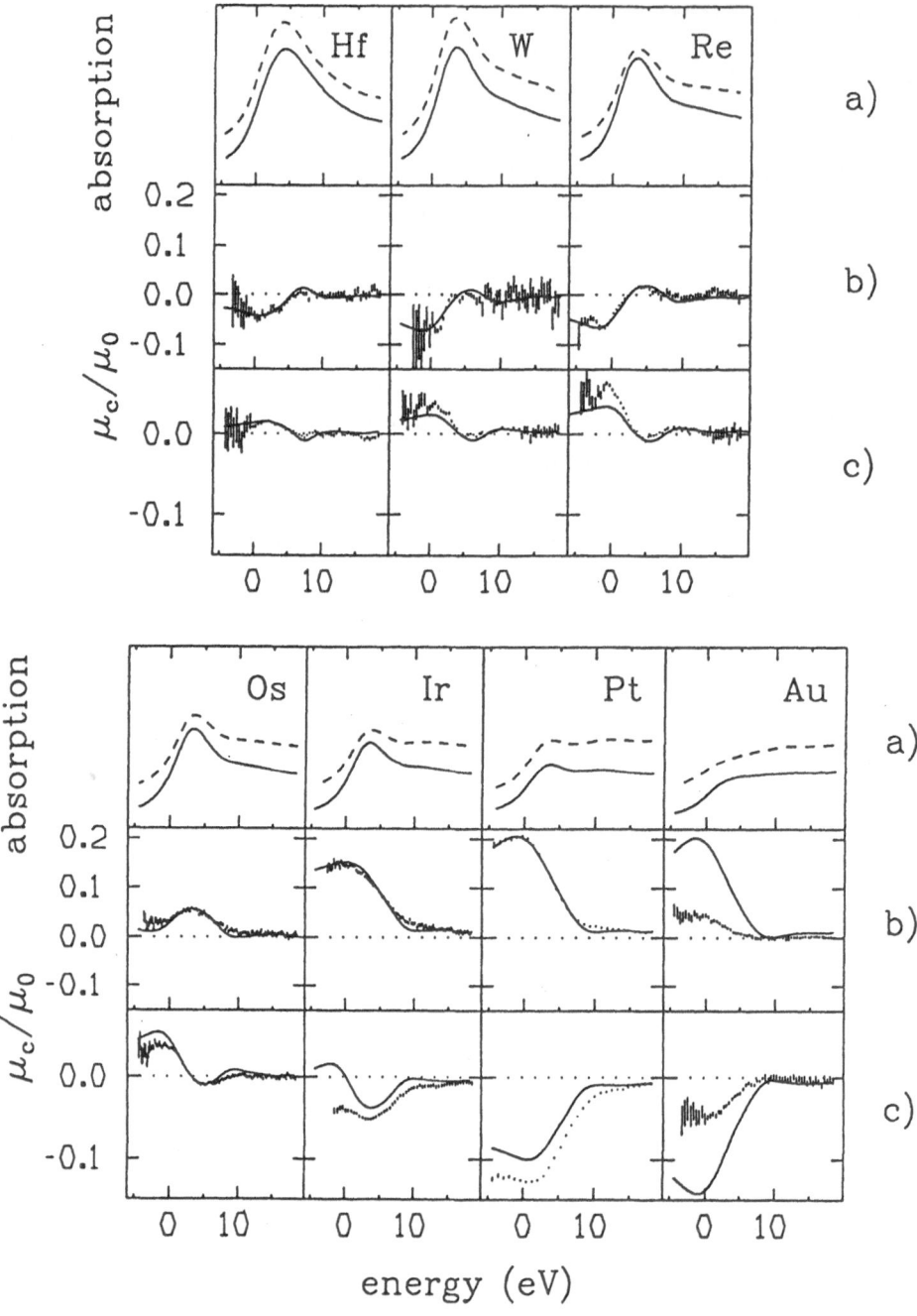

Fig. 3: Normal absorption profiles (a) at the L_2-edges (dashed line) and the L_3-edge (solid line) and spin-dependent (or CMXD-) spectra of the L_2- (b) and L_3- (c) edges of Hf-, W-, Re-, Os-, Ir-, Pt- and Au impurities in iron (3 at. %) in addition to theoretical calculations by H. Ebert et al. (solid lines in (b) and (c)).

3.2 5d impurities in Fe and a simplified picture of CMXD

Much larger spin-dependent absorption amounting to 20 % has been found at $L_{2,3}$-edges. The results for 5d-impurities in Fe (3 at.%) are presented in Fig. 3 in addition to the corresponding normal absoption profiles and the results of the theoretical calculations based on the same model as for the K-edges in iron [Ebe89, Ebe90a, Sch90b-c, Wie91]. Except for Au, where the energy dependence but not the amplitude is reproduced by the theoretical profiles, in all cases excellent agreement has been found between theory and experiment. It demonstrates, that this fully relativistic theory based on the single-particle band-structure picture allows a parameter-free description of the experimental spectra.

An interesting finding is, that except for Os the ratios between the CMXD at the L_2- and L_3-edges are in the range of $\mu_c/\mu_0(L_2)/\mu_c/\mu_0(L_3) \sim -1..-3$. This phenomenon, which has been also observed in the recent studies of the $L_{2,3}$-CMXD in 3d-elements [Chen90, Set90], as well as for Gd and Ce (shown in chapter 3.3) can be understood by simple theoretical considerations based on the «Fano» effect [Fan69, Ers75]. The physical origin of the spin-dependent near-edge absorption or CMXD is the transfer of the photon spin polarization to the photoelectron in the inner-shell absorption process as a consequence of angular momentum conservation and spin-orbit interaction of the initial and final state. Thus this photoelectron polarization P_e is small for K-absorption in an initial pure s-state $(P_e(K) \sim 1\%)$ but much larger for L_2- and L_3-absorption in the $2p_{1/2}$- and $2p_{3/2}$-states $(P_e(L_2) = -1/2$ and $P_e(L_3) = +1/4$ calculated for free atoms) due to their large initial-state spin-orbit splitting.

In a naive «two-step» model, which is scetched in Fig. 4 for a L_2-absorption process in a Pt impurity in PtFe, it is demonstrated that the photoelectron polarization causes a difference of the absorption coefficients for right (μ^+) and left (μ^-) circularly polarized light of $\mu^+ - \mu^- \sim P_e \cdot (\rho^+ - \rho^-)$. The normalized value

$$[\mu_c/\mu_0] (E) = P_e \cdot [\Delta\rho/\rho] (E) \tag{1}$$

is directly related to the spin density of $(\Delta\rho/\rho) = (\rho^+ - \rho^-)/(\rho^+ + \rho^-)$ of the final states, where $\rho\uparrow$ $(\rho\downarrow)$ are majority (minority) unoccupied bands. From eq. (1) it

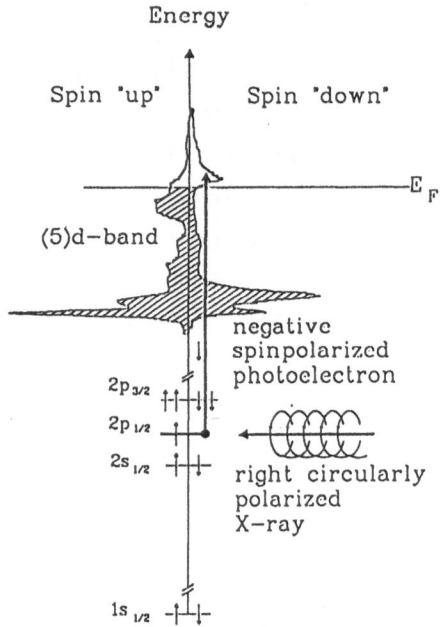

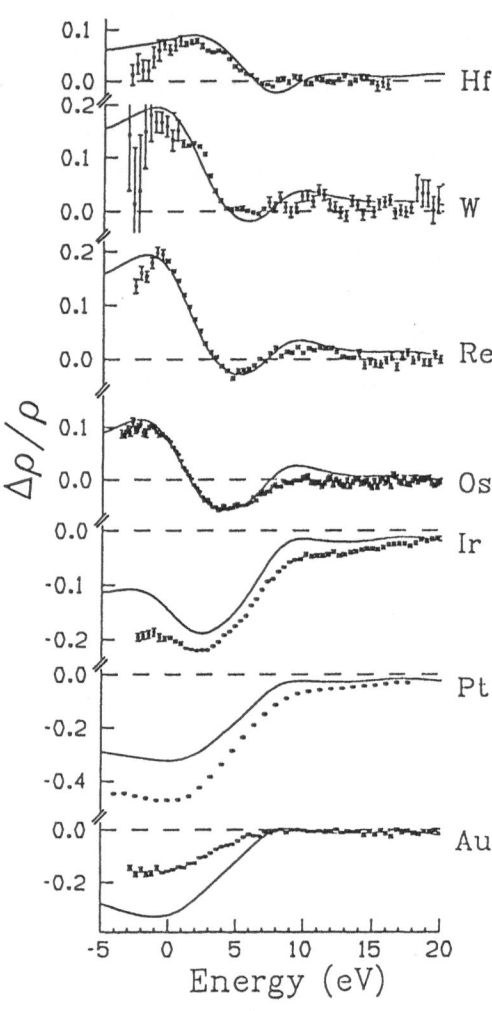

Fig. 4: Absorption of a right circularly polarized photon in a $2p_{1/2}$
level of an Pt atom oriented in
iron, whose spins are aligned parallel to the photon spin direction:
Due to the negative spin polarization of the photoelectron and the
larger final state density for spin-
down states $g^- > g^+$ the absorption
coefficient, described by $\mu^+ \sim g^-$,
is larger than for the absorption
of left circularly polarized x-rays
($\mu^- \sim g^+ < \mu^+$).

Fig. 5: Theoretical spin-density
profiles $\Delta g/g$ of d-like unoccupied
states of 5d impurities in iron
(solid line) and corresponding $\Delta g/g$
spectra deduced from the CMXD
signals (in Figs. 3).

can be well understood that in principle the dichroic effects at the K-edges have to be rather small due to the small value of P_e (K) ~ +0.01, which gives the upper limit of the CMXD at the K-edge. Since P_e is about 25 (50) times larger at L3- (L2 -) edges their magnetic absorption is much easier to detect. The ratio of P_e(L2)/P_e(L3) for free atoms is -2 leading to the same ratio for the CMXD $[\mu_c/\mu_o]$(L2)/$[\mu_c/\mu_o]$ (L3) ~ -2 if the spin density $[\Delta\rho/\rho]$ for the final states, into which the photoelectrons are transferred after L2- and L3-absorption, are similar.

However, due to the influence of spin-orbit interaction for L3-absorption the dipole matrixelements for the (2p3/2 → d5/2) - transition are much larger than for the other possible transitions (relative transition probabilities 2p3/2 → d5/2 (90%), 2p3/2 → d3/2 (9%), 2p1/2 → s1/2 (1%)). After L2-absorption about 98% of the photoelectrons populate a d3/2 empty state. Thus the final states after L2- and L3-absorption have different spin-orbit character. Although in case of a spin-dependent potential the total angular momentum is not a good quantum number, a deviation of the ratio -2 for L2- and L3-spin-dependent absorption profiles can be qualitatively understood as an influence of the final-state spin-orbit interaction on the spin-dependent absorption.

A comparison of the theoretical values for $\Delta\rho/\rho$ and the spin-density profile deduced from the CMXD signal by eq. (1) in Fig. 5 shows excellent agreement for almost all 5d-elements and confirms the validity of equation (1). A detailed theoretical study of the relation between the CMXD profile and the final-states spin polarization has been published by H. Ebert and R. Zeller [Ebe90b].

The d-band at the Fermi level of the 5d-impurities in iron is located in an energy region of about 20 eV. The integrated area of the density-of-states profiles correspond to approximately 5 spin-up and 5 spin-down electrons. A Pt atom carries a d-momentum given by the difference of the number of spin-up and spin-down d-bands in the occupied part, which has the same amplitude but opposite sign as the «hole»-momentum of the unoccupied d-band $m_d = - m_d^{hole}$ = $-\mu_B \cdot \int\Delta\rho$ (E) dE . Thus in the simplified picture described above the magnetic d-momentum

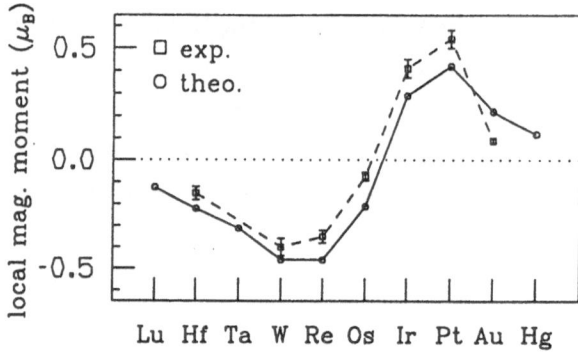

Fig. 6: Theoretical local magnetic d- moments of 5d impurities in iron and comparison with the experimental values deduced from the $L_{2,3}$- CMXD spectra.

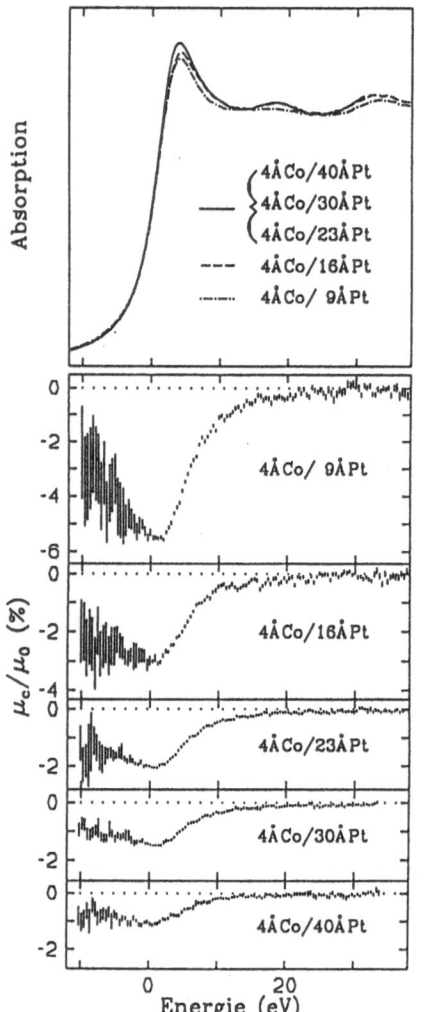

Fig. 7: Normal absorption and CMXD profile at the Pt L_3-edges in Pt/Co multilayers with different Pt thickness corresponding to 2 monolayers Co and 4,7,10,13 and 18 monolayers Pt.

Fig. 8: Theoretical (x) and experimental average Pt moment (a) and total Pt moment (b) in Co/Pt multilayers deduced from the spin-dependent absorption profiles in Fig. 7.

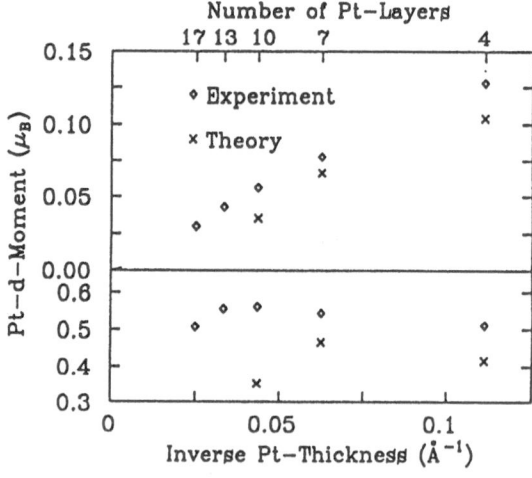

$$m_d = -\mu_B \, (1/P_e) \int [\mu_c/\mu_0] \, (E) \, \rho \, (E) \, dE \qquad\qquad (2)$$

can be directly deduced from the CMXD signal by integrating the spin-dependent absorption profile multiplied by the total density of the unoccupied states, which can be taken from theory or from measured white-line areas [Hor82, Sha87]. The experimental d-momenta [Sch90c, Wie91] determined by equation (2) and the theoretical d-momenta [Aka89, Ebe 90d] of 5d-impurities in iron are shown in Fig. 6. The theoretically predicted trend, especially the change from the antiferromagnetic to ferromagnetic coupling between Os and Ir, has been experimentally verified for the first time. The small deviations may result from systematic errors of the simplified pictures especially by taking a constant value of P_e [Ebe90b]. However, the good agreement with the calculations indicates, that spin-dependent absorption measurements can provide a fast and direct determination of the sign and qualitatively the amplitude of local magnetic moments, which cannot be gained for a very dilute system for example from neutron scattering studies.

An application of this method is demonstrated in case of Pt/Co multilayered structures (ML), which are discussed as new candidates for magneto-optical recording [Zep89] due to their large perpendicular anisotropy and coercivity. The role of the — in the pure metal non-magnetic — Pt interlayer for the magnetic properties of these artificial magnetic superlattices has been an open problem. The enhanced magnetic moment per Co atom in the ML was expected to result partially from a spin polarization or an induced magnetic moment of the Pt atoms [McGui84, Sch90a]. To shed light on this problem we have performed CMXD studies on several multilayered structures of different Pt thickness $30 \cdot (4 \, \text{Å Co} + X \, \text{Å Pt}, X = 9, 16, 23, 30, 40)$ [Rüe91] close to the optimum composition for magneto-optical recording of $30 \cdot (4 \, \text{Å Co} + 18 \, \text{Å Pt})$. Using the two beam transmission mode with PtFe (20 at %) as a reference target we have deduced from the spin-dependent absorption profiles shown in Fig. 7 the magnetic Pt d-moment as function of the Pt layer thickness with an accuracy smaller than 2%. The results compared with recent calculations of H. Ebert et al. [Ebe91a] are presented in Fig. 8. The experimental moments which for the first time prove, that the Pt interlayers carry about 15% of the total sample magnetization, have

Magnetic Properties

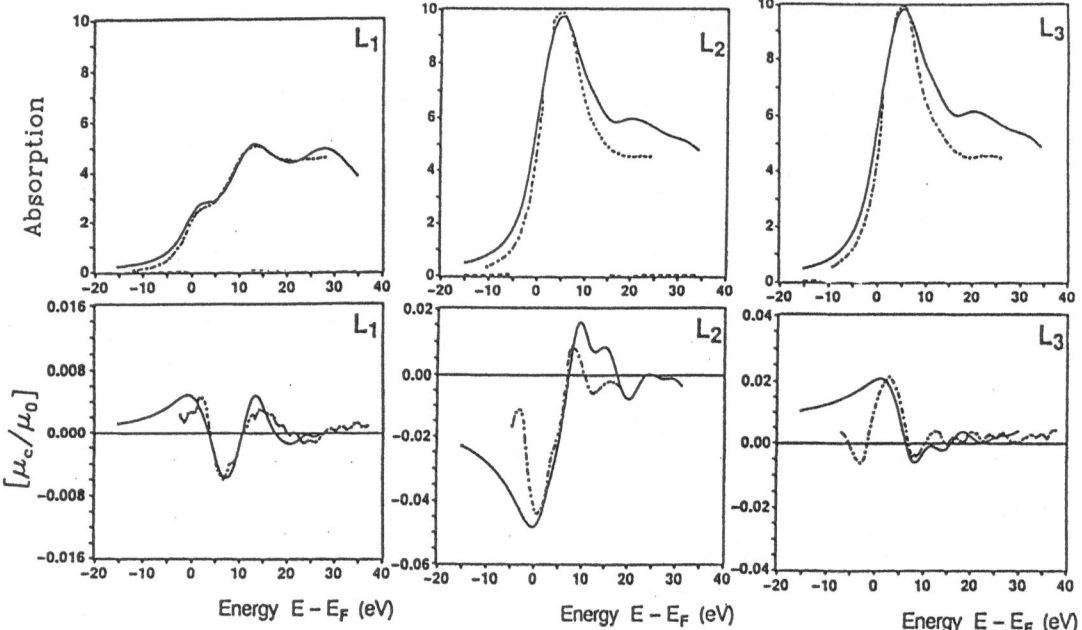

Fig. 9: Experimental (dashed line) and theoretical (solid line) Gd $L_{1,2,3}$-absorption and corresponding spin-dependent profiles.

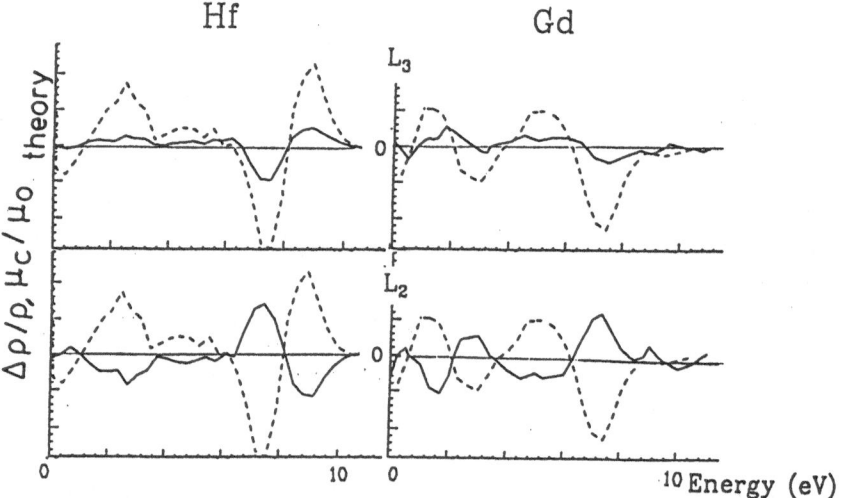

Fig. 10: Unbroadened unoccupied d-like spin density (dashed lines) and unbroadened L_3- (upper part) and L_2- (lower part) CMXD spectra calculated by H. Ebert for a Hf impurity in Fe and Gd in Gd metal.

somewhat larger amplitudes than expected from theory. Here one finds an increase of the total Pt moment for 7 monolayers Pt followed by a significant reduction in case of 10 monolayers Pt. It results from a calculated Pt spin polarization oscillating as function of the distance of the monolayer from the Co atoms with a small negative contribution of the inner Pt layers in case of the 10-monolayer Pt sample. The experimental findings on the total Pt d-moments show a similar trend. However, the oscillations indicated possibly by the CMXD studies have significantly smaller amplitudes and larger oscillation lengths.

3.3 CMXD in 4f-systems

The L-edges of 4f elementary ferromagnets or 4f atoms ferro(i)magnetically oriented in 3d-4f systems and oxides show also large CMXD signals with amplitudes of some percent. It was experimentally verified for a lot of rare-earth elements as Ce, Nd, Sm, Gd, Tb, Ho, that the gross structure of the spin-dependence of the $L_{2,3}$-absorption profile is only weakly influenced by the chemical environment [Sch89b, Sch89e, Sch90b, Fis90, Bau90a]. Nevertheless, the interpretation of the 4f-CMXD is much more complicated than for pure spin systems as the 5d-impurities in iron. A comparison of the d-like spin-densities of Gd metal [Sti85] and the spin-density profiles for Gd metal deduced from the first 4f $L_{2,3}$-CMXD spectra by eq. (1) [Sch88] indicated that there exists no simple relation between the spin polarization of states at the Fermi level or magnetic d-moments for the 5d-elements.

Recently L-CMXD calculations by H. Ebert et al. [Ebe90c] and P. Carra et al. [Car90, Car91] based on the single-particle band-structure picture have become available. As seen from Fig. 9 they describe the experimental data well in case of the L_1-edge and the $L_{2,3}$-absorption for positive energies. A comparison of the unbroadened theoretical μ_c/μ_o-spectra and the (unbroadened) theoretical spin-density profile $\Delta\rho/\rho$ calculated by H. Ebert for Hf and Gd shown in Fig. 10 indicate, that relation (1) gives a simple interpretation in case of the lighter 5d-elements as Hf (atomic configuration $4f^{14}\,5d^2\,6s^2$), since the ratio of $[\mu_c/\mu_o]\,(E)\,/[\Delta\rho/\rho]\,(E) = P_e\,(E)$ corresponds to the constant free-atom value of the photoelectron polarization $P_e\,(L_3) = +\,1/4$ and $P_e\,(L_2) = -\,1/2$. But for Gd (atomic configuration $4f^7 5d^1 6s^2$) P_e is significantly energy dependent with strongly increased values amounting to P_e (L2) ~ -1 and P_e (L3) ~ +0.5 at the Fermi level E = 0.

The experimental L2, 3-spectra in Fig. 9 show a significant decrease of the L2, 3-CMXD signal at the lower-energy side of the white line especially for the L3-edge, which is not found in the theoretical profile. This discrepancy was explained by P. Carra and coworkers as negative quadrupole contributions to the spin-depen-dent profile resulting from atomic E2-transitions to the empty, in case of Gd negative spin polarized, 4f-state. The strength of these energetically low lying E2-transitions should in case of 4f-elements amount to some percent relative to the dominant (2p → d) E1-transition and have been clearly observed in the magnetic resonant scattering mode [Han88]. The features of the CMXD spectrum associated with a E2-contribution can show a different dependence on the angle Θ between the magnetization of the absorbing atom and the photon beam direction than the E1-part, which varies with cos Θ. In this case it may be experimentally possible to identify the multipole character of the features of the μ_c/μ_o-profile. First experimental studies on Gd and Er show no deviation from the cos Θ dependence within the statistical accuracy [Fis91]. We hope that further investigations can clarify this important point, since the dipole and quadrupole parts in the CMXD spectra contain information on the spin polarization of the empty localized 4f states and the outer d-bands separately.

In 4f-systems quadrupole transitions can contribute and a strong energy depen-dence of the relation between the final-state spin-polarization and the L2, 3-CMXD profile exists. Due to these complications, its interpretation in terms of spin densities and local magnetic moments is only possible in connection with reliable theories.

However, from an experimental study of the L2, 3-spin dependent absorption in an almost closed RE row (RE = Ce, Pr, Nd, Sm, Gd, Tb, Ho, Er and Tm) of the binary metallic compound $(RE)_2Co_{17}$ shown in Fig. 11 some systematics can be deduced. Therefore the L2- and L3-CMXD spectra $(\mu_c/\mu_o)^{L3} \cdot (+4)$ and $(\mu_c/\mu_o)^{L2}$ (-2) have been rescaled by factor $1/P_e$ to take into account the free atom

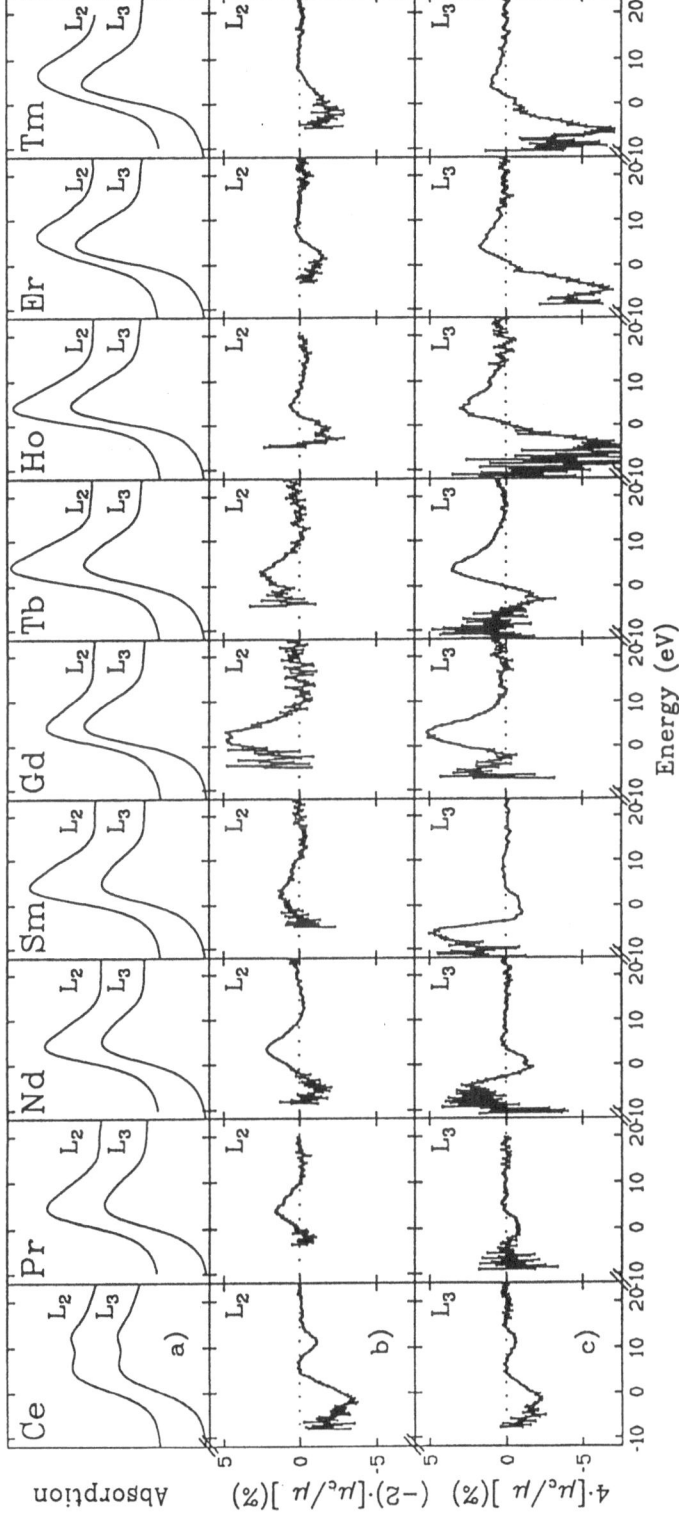

Fig. 11: Normal L_2- and L_3-absorption profiles and rescaled spin-dependent absorption spectra $[\mu_c/\mu]/P_e$ (not normalized for complete 4f orientation) at the L_2- (b) and L_3- (c) edges of the rare-earth atoms in $(RE)_2Co_{17}$ relative to the direction of the RE *spin* lattice.

photoelectron polarization. In this context a *positive* sign of the rescaled μ_c/μ_0-spectra is interpreted in terms of a spin-orientation of the populated final state parallel to the 4f spins, which are *parallel* to the spins of the rare-earth (5)d electrons and *antiparallel* to the Co spin sublattice. It is obvious, that a strong variation exists of the $L_{2,3}$-CMXD spectra within the 4f row. In addition, large differences between the rescaled L_2- (Figs. 11b) and L_3- spectra (Figs. 11c) of the same 4f-element have been found except for Gd and Ce, where the rescaled L_2- and L_3-spectra have nearly equal gross structures and amplitudes. Tb shows a somewhat intermediate behaviour between Gd and the heavier rare earth concerning the similarity of the $(\mu_c/\mu_0)^{L_2} \cdot (-2)$ and $(\mu_c/\mu_0)^{L_3} \cdot (+4)$ spectra. Since the Gd^{3+} ion is in an ionic s-state and the 4f angular momentum of Tb^{3+} (L=3) is smaller compared to the other RE's Pr, Nd, Sm, Ho, Er and Tm with L=5 or 6 these similarity (or a ratio of the CMXD signals of $(\mu_c/\mu_0)^{L_2}/(\mu_c/\mu_0)^{L_3} \sim -2$) seems to be connected with a small angular momentum of the 4f-core. In Pr, Nd, Sm, Ho, Er and Tm (L=5, 6) no simple relation between the L_2- and L_3-CMXD can be deduced, possibly indicating a strong influence of final-state spin-orbit effects on the spin-dependent absorption induced by the 4f-core angular momentum.

The normal $L_{2,3}$-absorption of Ce in Ce_2Co_{17} shows a double peak structure corresponding to two energetically shifted final states, which are typical for a variety of intermetallic compounds and insulators [Woh84, Röh87, Kai88, Fin81]. Whether this feature can be ascribed in the single-particle picture to a mixed valency in the ground state or to core-hole induced many-body effects has given rise to lively discussions. In the CMXD studies for both involved final states very similar spin dependence is found. That the ratio of the rescaled $L_{2,3}$-CMXD spectra is for both peaks also close to one, as found in Gd, might be an indication of a small or quenched angular momentum of the 4f-core in Ce_2Co_{17}. This assumption seems to be in agreement with recent calculations for the familiar systems $CeCo_5$ [Nor90] and $CeFe_2$ [Eri88], which show very similar normal and spin-dependent absorption profiles [Fin84, Bau90a]. In context of the systematics of the 4f $L_{2,3}$-CMXD discussed above, for both final states the 4f-angular momentum should have similar small values, a finding, which might be important for a better understanding of these double-peak structures.

Similar to the coupling mechanism of the lighter 5d-impurities in 3d-hosts, as expected from theory [Eri88, Nor90] and experimental studies, for all RE elements inclusive Ce the 5d-spin momentum (m^d (Ce) ~ -0.2 μ_B) and the 4f-spin momentum couples antiparallel to the Co spins [Ken90]. Thus the d-like spin density, near the Fermi level in Ce should have the same sign as Gd ($|m^d|$ ~ 0.5 μ_B calculated for Gd metal [Sti85, Tem90]). This is in striking disagreement with the completely different signs of the Gd and Ce μ_c/μ_o-profiles shown in Fig. 11. A comparison of the rescaled Ce-CMXD spectra with the corresponding Gd- and Hf-data of HfFe shown in Fig. 12 demonstrates, that the CMXD of Ce and Hf are very similar reflecting the correct negative spin polarization of the empty Ce-(5)d states, while due to the strong energy dependence of P_e (see Fig. 10) Gd behaves extraordinarly.

Thus as for Hf in HfFe an interpretation of the Ce μ_c/μ_o-spectra by eq. (2) in terms of local magnetic moments might be possible. It leads to a Ce (5)d-momentum of about -0.2 μ_B for the both final states corresponding to the first peak and second peak at higher energies. This reasonable agreement with the theoretical expectations for Ce in CeCo5 (m^d = -0.2 μ_B) might confirm the assumption, that for itinerant spin ferromagnets as the 5d-3d dilute alloys, as well as the Ce-3d metallic compounds, the simplified picture leading to eq. (1) and (2) is valid and a simple interpretation of the $L_{2,3}$-CMXD is possible.

4 Summary

In the last years the spin-dependent absorption spectroscopy has been developed as a new tool to study magnetism with synchrotron radiation. Newly developed theories allow in a lot of cases a parameter-free description of the observed phenomena confirming the validity of the underlying single-particle band-structure picture. Since in a very simplified model the spin-dependent profile can often be directly interpreted in terms of local spin density distributions and local magnetic moments the element-selective CMXD spectroscopy is an effective experimental method to get important information on local magnetic structures even for materials, for which no theories are available especially for complex multicomponent systems.

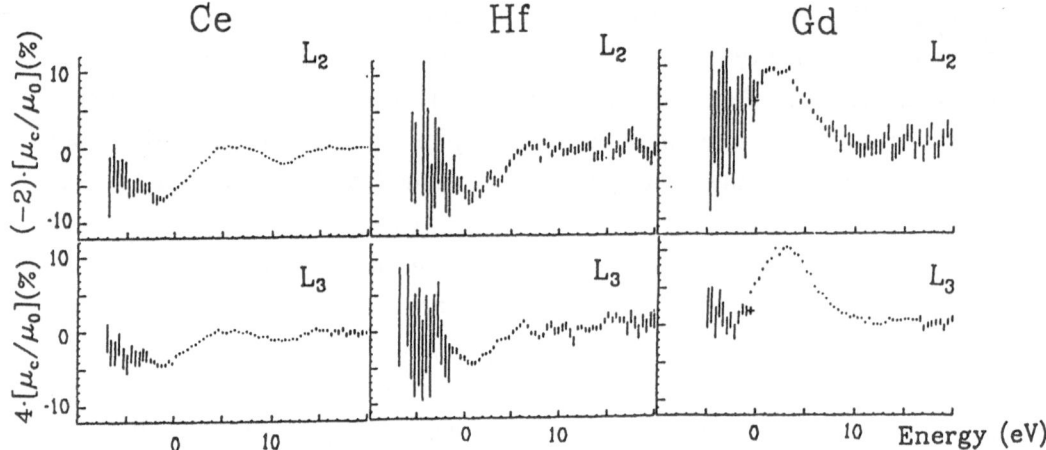

Fig. 12: Rescaled Ce CMXD $L_{2,3}$-spectra of Ce_2Co_{17} (same as in Fig. 11 for complete 4f orientation) in comparison with the corresponding spectra of Hf of Hf<u>Fe</u> and Gd in Gd_2Co_{17} to the 5d spin direction.

References

[Aka89]   Akai H., Hyp. Int. 43 255 (1989).

[Bau90a] Baudelet F., Dartyge E., Krill G., Kappler J.P., Brouder C., Piechuch
          M. and Fontaine A., Phys. Rev. B (1991) in press.

[Bau90b] Baudelet F., Brouder C., Dartyge E., Fontaine A., Kappler J.P. and Krill
          G., Eur. Phys. Lett. (1991) in press.

[Cam66]   Campbell I.A., Proc. Phys. Soc. 89 71 (1966)

[Car89]   Carra P., Altarelli M. and de Bergevin F. Phys. Rev. B40 7324 (1989)

[Car90]   Carra P. and Altarelli M., Phys. Rev. Lett. 64 1286 (1990).

[Car91]   Carra P., Harmon B.N., Thole B.T., Altarelli M. and Sawatzky G.A. to be
          published.

[Chen90] Chen C.T., Sette F., Ma Y. and Modesty S., Phys. Rev. B42 7262 (1990).

[Cla91]   Clarke R., Lamelas F.J., Hui H.D., Baudelet F., Dartyge E. and Fontaine
          A., J. Magn. Magn. Mat. (1991) in press.

[Col89]   Collins S.P., Cooper M.J., Brahmia A., Laundy D. and Pitkanen T.J., Phys.
          Cond. Mat. 1 323 (1989).

[Ebe88a] Ebert H., Strange P. and Gyorffy B.L., J. Appl. Phys. 63 3055 (1988).

[Ebe88b] Ebert H., Strange P. and Gyorffy B.L., Z. Phys. B73 77 (1988).

[Ebe89]   Ebert H., Drittler B., Zeller R. and Schütz G., Sol. Stat. Comm. 69 485 (1989).

[Ebe90a] Ebert H., Wienke R., Schütz G. and Zeller R., J. Appl. Phys. 67 4923
          (1990).

[Ebe90b] Ebert H. and Zeller R., Phys. Rev B42 2744 (1990).

[Ebe90c] Ebert H., Schütz G., Temmerman W.M., Solid State Comm. 76 475 (1990).

[Ebe90d] Ebert H., Wienke R., Schütz G. and Temmerman W.M., Physica A (1991)
          in press.

[Ebe90e] Ebert H., Zeller R., Drittler B. and Dederichs P.H., J. Appl. Phys. 67 4576
(1990).

[Ebe91a] Ebert H., Rüegg S., Schütz G., Wienke R. and Zeper W.B., J. Magn. Magn.
          Mat. (1991) in press.

[Ebe91b] Ebert H. and Schütz G., J. Appl. Phys. (1991) in press.

[Eri88]   Ericson O., Nordström L., Brooks M.S.S. and Johansson B., Phys. Rev. Lett.
          24 2523 (1988).

[Ers75]   Erskine J.L. and Stern E.A., Phys. Rev. B12 5016 (1975).

[Dar86]   Dartyge E., Depautex C., Dubuisson J.M., Fontaine A., Jucha A.,
          Leboucher P. and Tourillon G., Nucl. Instr. Meth. A246 452 (1986).

[Fan69]   Fano U., Phys. Rev. 178, 131 (1969).

[Fin81]   Finkel'shteyn L.D. and Samsonova S.D., Phys. Met. Metall. 53 84 (1981).

[Fis90]   Fischer P., Schütz G. and Wiesinger G., Solid State Comm. 76 777 (1990).

[Fis91]   Fischer P., Schütz G. and Wiesinger G., J. Appl. Phys. (1991) in press.

[Han88]   Hannon J.P., Trammel G.T., Blume M. and Gibbs D., Phys. Rev. Lett. 61
            1245 (1988).

[Hor82]   Horsley J.A., J. Chem. Phys. 79 1451 (1982).

[Kai88]   Kaindl G., Schmiester G., Sampathkumaran E.V. and Wachter P., Phys. Rev.
            B38 10174 (1988).

[Ken90]   Kennedy S.J. and Coles B.R., J. Phys. Cond. Matt. 2 1213 (1990).

[Kes74]   Keski-Rahkonen O., Krause M.O., At. Data Nucl. Data Tables 14 139 (1974).

[Laa86]   van der Laan G., Thole B.T., Sawatzky G.A., Goedkoop J.B., Fuggle J.C.,
            Esteva J.-M., Karnatak R., Remeika J.P. and Dabkowska H.A., Phys. Rev.
            B34 6529 (1986).

[McGui84]   McGuire T.R., Aboaf J.A., Klokholm E., J. Appl. Phys. 55 1951 (1984).

[Nor90]   Nordström L., Ericson O., Brooks M.S.S. and Johansson B., Phys. Rev. B41
            9111 (1990).

[Pfl90a]   Pflüger J. and Heintze G., Nucl. Instr. Meth. A246 300 (1990).

[Pfl90b]   Pflüger J., Heintze G., Frahm R., Wienke R., Wilhelm W., Fischer P. and
            Schütz G., Conf. Proc. Vol. 25 «2nd European Conf. on Progress in X-ray
            Synchrotron Radiation Research» A. Balerna, E. Bernieri and S. Mobilio (Eds.)
            SIF, Bologna (1990).

[Röh87]   Röhler J. in «Handbook on the Physics and Chemistry of Rare Earths, Vol.
            10» ed. by K.A. Gschneidner jr., L. Eyring and S. Hüfner (North Holland 1987).

[Rüe91]   Rüegg S., Schütz G., Fischer P., Wienke R., Zeper W.B. and Ebert
            H., J. Appl. Phys. (1991) in press.

[Sch87]   Schütz G., Wagner W., Wilhelm W., Kienle P., Zeller R., Frahm R. and
            Materlik G., Phys. Rev. Lett. 58 737 (1987).

[Sch88]   Schütz G., Knülle M., Wienke R., Wilhelm W., Wagner W., Kienle P. and
            Frahm R., Z. Phys. B73 67 (1988).

[Sch89a]   Schütz G., Wienke R., Wilhelm W., Wagner W., Kienle P., Zeller R. and
            Frahm R., Z. Phys. B75 495 (1989).

[Sch89b]   Schütz G. and Wienke R., Hyp. Int. 50 457 (1989).

[Sch89c]   Schütz G., Physica Scripta, T29 172 (1989).

[Sch89d]   Schütz G., Wienke R., Wilhelm W., Wagner W., KienleP., Zeller R. and
            Frahm R., Physica B158 284 (1989).

[Sch89e]   Schütz G., Frahm R., Wienke R., Wilhelm W., Wagner W. and Kienle P.,
            Rev. Sci. Instr. 60 1661 (1989).

[Sch90a]   Schütz G., Wienke R., Wilhelm W., Zeper W.B., Ebert H. and Spörl K.,
            J. Appl. Phys. $\underline{67}$ 4456 (1990).
[Sch90b]   Schütz G., Fischer P., Ebert H., Wienke R. and Wilhelm W., Conf. Proc.
            Vol. 25 «2nd European Conf. on Progress in X-ray Synchrotron Radiation
            Research» A. Balerna, E. Bernieri and S. Mobilio (Eds.) SIF, Bologna (1990) S. 229.
[Sch90c]   Schütz G., Physikalische Blätter $\underline{46}$ 475 (1990).
[Set90]   Sette F., Chen C.T., Ma Y., Modesti S. and Smith N.V., Conf. Proceedings
            «6th Int. Conf. on X-ray Absorption Fine Structure» (Ellis Horwood LTD, 1990).
[Sha87]   Sham T.K., Solid State Comm. $\underline{64}$ 1103 (1987).
[Sti85]   Sticht J. and Kübler J., Solid State Comm. $\underline{53}$ 529 (1985).
[Tem90]   Temmerman W.M. and Sterne P.A., J. Phys. Cond. Mat. $\underline{2}$ 5529 (1990).
[Tho85]   Thole B.T., van der Laan G. and Sawatzky G.A., Phys. Rev. Lett. $\underline{55}$ 2086
            (1985).
[Wie91]   Wienke R., Schütz G. and Ebert H., J. Appl. Phys. (1990) in press.
[Woh84]   Wohlleben D. and Röhler J., J. Appl. Phys. $\underline{55}$ 1904 (1984).
[Yam89]   Yamamoto S., Kawata H., Kitamura N., Phys. Rev. Lett. $\underline{62}$ 23 (1989).
[Zep89]   Zeper W.B., Greidanus F.J.A.M., Carcia P.F. and Fincher C.R., J. Appl. Phys.
            $\underline{61}$ 4971 (1989).

Synchrotron Radiation Spectroscopies:
Spin Polarized Electron Yield
in X-ray Absorption

F.Sirotti[1], S.Toscano[2], A.Waldhauer[1], and G.Rossi[1,2]

[1] L.U.R.E. Université de Paris Sud, F-91405 Orsay
[2] Laboratorium fur Festkoerperphysik ETH, CH-8093 Zurich

Introduction

The experimental investigation of surfaces and interfaces with spectroscopy aims to the characterization of microscopic atomic and electronic order parameters which may clarify the fundamental implications of such macroscopic behaviours as contact potentials, Schottky barriers, surface magnetism, work functions. Synchrotron radiation methods for surface science allow the definition of local and extended probes of the atomic order (SEXAFS and Photoelectron Diffraction vs. Grazing Incidence X-ray Surface Diffraction and X-standing Wave Spectroscopy) and of the electron states (hν-dependent photoemission, core level spectroscopy, X-ray Absorption spectroscopy, Angle Resolved Photoemission). In the domain of surface magnetism the development of local probes is also underway with the Spin-resolved Auger Spectroscopy[1], the recent first results of Spin-resolved core level photoemission (or SPESCA)[2] to complement the Angle and Spin resolved PES of single crystal valence bands, or the magnetometric measures of the spin polarization of the total electron yield[3].

In this paper we present the synchrotron radiation spectroscopy station SU7 of LURE which we recently developed by adding a spin detector and which represents a powerful tool for spectroscopy in the soft X-ray regime.

We also present in the final section preliminary data on $L_{2,3}$ edge X-ray absorption with spin analysis of the secondary electron Yield for polycrystalline Fe and Co films.

In order to improve the understanding of the properties of a material, it is necessary to perform measurements using different techniques. The most reliable results are obtained when the same sample is used during the various measurements, and moreover, when the sample remains in the same environment during the period of investigation. This second point is especially important when surfaces are considered, since these are the most exposed to external influences. The sensitivity to external parameters increases when one considers effects which involve small changes in energy, as in the case of magnetism.

For this motivation, an electron spin analyser was added to the beam line XPS-SU7 at the synchrotron radiation source SuperAco.

Beam line description:

The experimental set up shown in the schematic diagram of Fig. 1 allows one to perform a complete surface characterization with energy dependent spectroscopies and magnetometry with electrons on the same sample in ultra high vacuum (P< $1 \cdot 10^{-10}$ mbar). The instrumentation control for the measurements and the modification of the sample parameters (temperature and magnetic field) are done by a personal computer with a GP-IB interface. Some considerations on the program to perform the measurements are given after the description of the corresponding instruments. In addition two programs are available, one to adjust the monochromator, i.e. zero order determination and 'manual' definition of the photon energy, and the other to plot previously acquired data files.

The XPS chamber has a lower level for sample preparation with an ion gun for sputtering and 3 or more evaporators for film deposition. Deposition rate is monitored by a quartz microbalance with a resolution of .1 Angstrom. The sample holder designed for magnetometry experiments allows the application of a variable magnetic field created by a horseshoe electromagnet and the variation of the sample temperature in the range 10K-300K. The maximum possible intensity of the magnetic field on the sample is 80Gauss with a current of 10 amperes. Also the current source is GP-IB interfaced for remote operations, in order to execute automatic measurements. The measurements are performed at L.U.R.E. with the Superaco synchrotron radiation source on the line

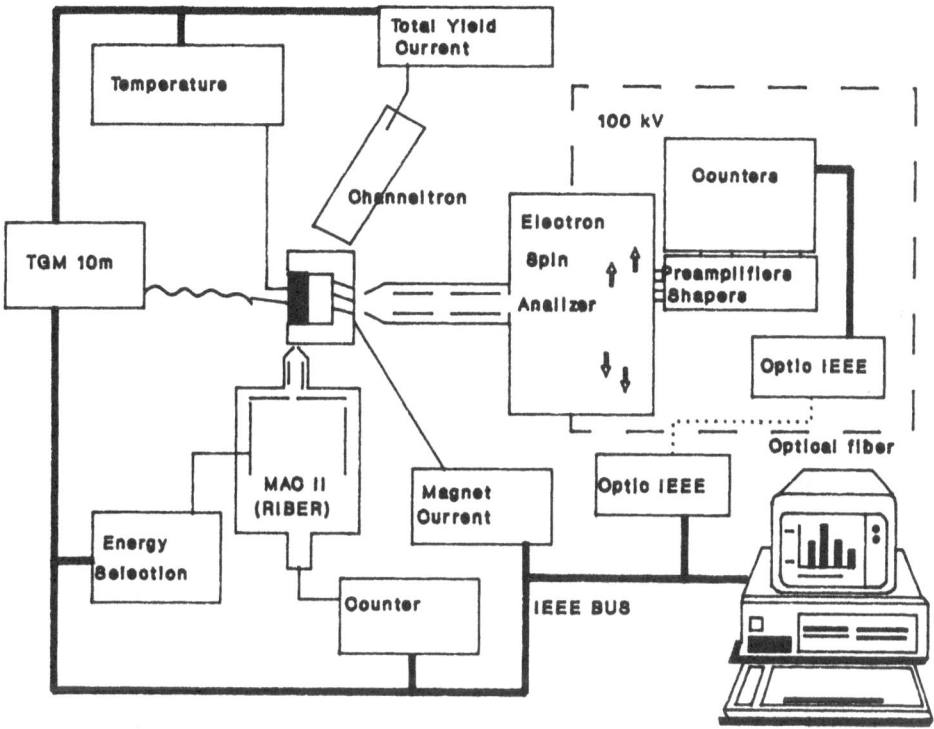

Fig. 1: Schematic diagram of the experimental set-up at the beam line SU7 of the SuperAco synchrotron radiation source.

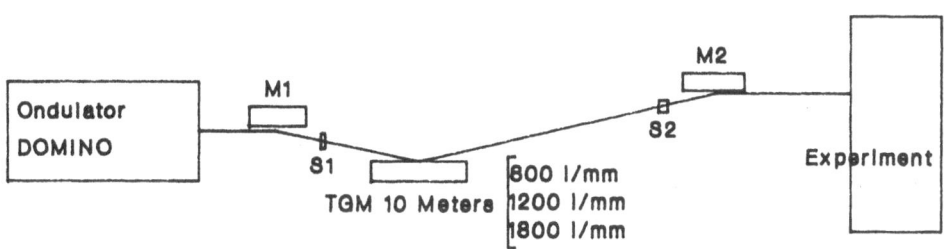

Fig. 2: Schematic diagram of the beam line SU7 at L.U.R.E.

SU7 which accepts radiation from the undulator Domino[4]. The line is equipped with a toroidal grating monochromator (TGM) with three different gratings of 800, 1200 and 1800 lines/mm and covers the energy range 200–800 eV. The schematic diagram of the beam line is shown in Fig. 2; the synchrotron radiation produced by the undulator is focused by a toroidal mirror (M1) and collimated by the entrance slit (S1). The monochromatic radiation is collimated by the exit slit (S2) and again focused on the sample by the second toroidal mirror (M2). The monochromatic radiation is defined by tilting the selected grating with a stepping motor interfaced to a personal computer. The transmission spectra of the two gratings used during this experiment are shown in Fig.3. The spectra are normalised to the SuperAco current and were collected with the same slit width of 200 um. To evaluate the resolution of the beam line at the Fe-L_3-edge we have compared the XAS spectrum with a broadened high resolution electron energy loss spectrum[5]. The broadening necessary to fit the two spectra corresponds to a energy resolution of 1eV for the spectrum obtained with an entrance slit of 40um and an exit slit of 60um.

The presence of the electron energy analyser [6] with the tunable monochromatic synchrotron radiation source allows one to perform a wide range of energy dependent spectroscopies as Energy Dispersion Curves (EDC), Constant Final State (CFS), Constant Initial state (CIS) Auger electron spectroscopy (AES). The EDC measurement is done fixing a primary photon energy and then analysing the energy of the emitted electrons, CIS and CFS spectra are obtained scanning photon energy with the monochromator, in the first case the electrons of constant energy are measured while in CFS is kept constant the difference between the energy of the electrons and of the photons.

The pulse signal from the high current channeltron located at the end of the electrons energy analyser is collected with a counter (GP-IP interface)[7]. AES measurements are done with a lockin[8] and the primary electron beam of an electron gun. The channeltron shown in Fig. 1 allows Total Yield measurements in current mode. Low Energy Electron Diffraction (LEED) is used to evaluate surface structure.

Magnetometry with electrons:

The low energy secondary electrons emitted from the sample are imaged with an electron lens system on the entrance diaphragm of a 100 kV Mott detector (ETH Zurich). The high energy Mott scattering [9] is used to measure electron-spin polarization. The left-right scattering asymmetry produced as a result of spin-orbit coupling is measured when high-energy electrons are scattered at a large angles from the nuclei in a thin gold foil. The elastically scattered electrons are detected at a scattering angle of 120° by four solid state detectors in the up-down and left-right direction.

The signal is integrated by a Canberra silicon surface barrier detector preamplifier [10]. After differentiation, shaping, and discrimination the pulses are counted by four counters [11] (GP-IB interface). All this apparatus operates at 100 kV potential. Instrumentation control and data acquisition is performed using a GP-IB optical interface[12] for the connection with the personal computer. The presence of the four detectors allows two different orientations of the magnetic field on the sample and/or the control of the correct alignment of the optic. The spin-polarization is defined as $P = (N\uparrow - N\downarrow)/(N\uparrow + N\downarrow)$ where $N\uparrow(N\downarrow)$ is the number of electrons with magnetic moment parallel (antiparallel) to the sample magnetization M. The effective asymmetry (Sherman) function[9] of the analyser S_{eff} is 0.22. The particular program developed for this instrument is able to perform data acquisition as a function of different variables:

- Magnetic field, in this case a hysteresis loop of the sample is obtained. In Fig. 4 is reported the hysteresis loop of 20 Åof Fe on Permalloy substrate for the proper polarization direction (Left/Right) and the direction parallel to the magnetic field (Up/down). The polarization is calculated referring to the asymmetry $Q(0)$ measured on the sample with no field applied:

$$P(H) = \frac{100}{S_{eff}} \cdot \frac{Q(H) - Q(0)}{Q(H) + Q(0)}$$

- Photon energy, in this case two different experimental procedures are available, the remanence and saturation measurements. In the first case the spin polarization is measured for an applied magnetic field in each (opposite) direction, which gives the remanence polarization of the sample:

$$P(H) = \frac{100}{S_{eff}} \cdot \frac{Q\uparrow - Q\downarrow}{Q\uparrow + Q\downarrow}$$

In the second case the measurement is done with a defined magnetic field applied during the detection of the spin asymmetry and referring the polarization calculation to the asymmetry measured for the first point. For each point the temperature of the sample is recorded in order to determine the polarization of the sample versus temperature without changing other parameters or to monitor the temperature behaviour during the other measurements (i.e. heating of the sample due to the current for the magnetic field).

$L_{2,3}$-edge X-ray absorption with spin analysis of the secondary electron yield for polycrystalline Fe and Co films:

We have performed for the first time measurements of the absorption coefficient of Fe and Co in the total electron yield mode with spin resolution of the secondary yield. The spin polarization of the yield of true secondary electrons[13] bears a proportionality to the surface and bulk magnetization, but with a polarization enhancements with respect to the true magnetization[14]. The polarization of electrons of higher final state energy with respect to the cascade reflects, on the other hand, the band polarization.

By studying the excitation energy dependence of the spin polarization one expect to learn more about the relationship between polarization and magnetization. By varying the photon energy in a photoemission experiment one probes the solid with variable sensitivities to the states of different orbital character due to the energy dependence of the photoionization cross sections, and to the presence of Cooper minima. When the photon energy reaches the excitation threshold for a core level resonant transitions to the empty states just above the Fermi level may occur, according to the dipole selection rules, and to the band structure of the solid. By XAS in the soft X-ray range one can probe the empty 3d band states of the transition metals by exciting the 2p core electrons at threshold. In this case dipole transitions obey J→J+1 and the p→d channel dominates by an order of magnitude the p→s channel.

One can measure the X-ray absorption in the total electron yield mode i.e by detecting all the electrons produced by the the decay of the core holes created by the incident X-rays, and their secondaries. This means that the yield is due to Auger electrons, photoelectrons and the secondary cascade. At threshold no elastic photoelectrons arising from the new transition can escape into vacuum since a kinetic energy larger than the work function of the solid is required and therefore the total electron yield is due to the yield excited at below threshold energy (photoelectrons, and Auger of external shells and secondaries) plus the new Auger electrons due to the decay of the core hole and their secondary cascade.

If the core holes created are spin polarized then the polarization of the total yield will reflect this core polarization via the addition of spin polarized Auger electrons to

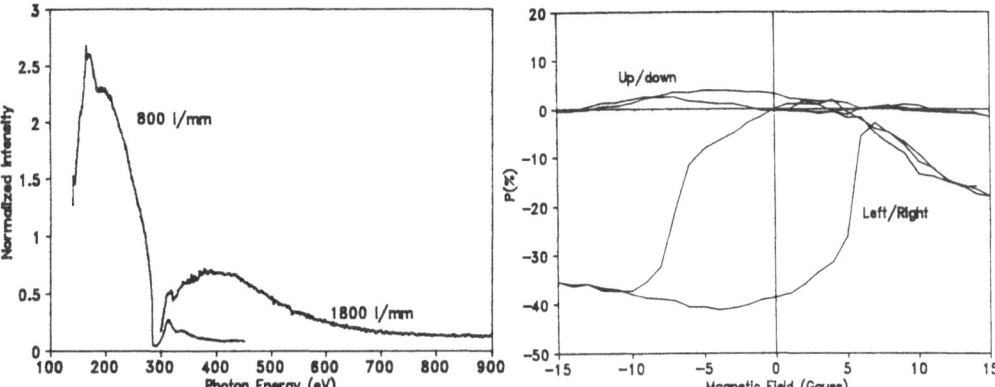

Fig. 3: Transmission spectra of the two gratings of 800 and 1800 lines/mm of the TGM monochromator.

Fig. 4: Hysteresis loop measured on 20 Åof Fe evaporated on a permalloy substrate. Data for asymmetry in the proper polarization direction (left/right) and in the direction parallel to the magnetic field (up/down) are reported.

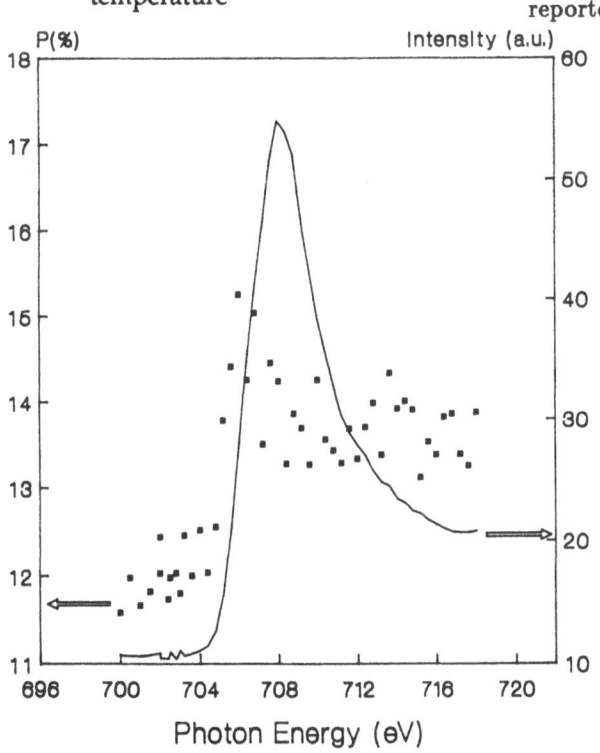

Fig. 5: XAS intensity and spin polarization L_3-edges obtained for a polycrystalline Fe film.

the total yield. We note that in Auger decay the ejected electron is polarized opposite to polarization of the initial hole[15,16]

In bcc-Fe the empty d-band states are basically all of minority spin. Once the spin quantization axis is defined by applying a magnetic field parallel to the surface of the sample and aligned with the detectors of the Mott scatterer one can therefore measure the spin of the total yield when exciting the Fe sample with monochromatic energies reaching the $2p_{\frac{3}{2}}$ threshold. The variation obtained in the spin polarization on the edge with respect to below the edge will therefore be a measure of the empty band spin polarization.

We show in figure 5 the results of typical XAS intensity and polarization edges obtained on polycrystalline Fe films. The intensity curve is just a measure of the X-ray absorption of Fe $2p_{\frac{3}{2}}$, it shows a prominent peak representing the width of the empty part of the Fe 3d band in presence of a 2p core hole. Its intensity and width represent the magnitude and energy distribution of the empty 3d states.

The spectrum is measured with a photon energy resolution of the order of one electron-volt. The polarization of the total yield also displays a characteristic edge at the threshold energy for Fe $2p_{\frac{3}{2}}$ (and Co $2p_{\frac{3}{2}}$). The spin polarization increases by 3% when crossing the edge and falls at higher energies to a plateau roughly 1% higher than below edge. This result can be understood in terms of spin polarization of the 2p core holes. The absorption 2p→3d can be seen as a change of a 3d valence hole with a 2p core hole and therefore, due to the minority spin character of all 3d holes only 2p down-spin holes can be created at threshold. This is reflected in an increase in polarization due to the up-spin polarized electrons ejected in the Auger decay of the 2p holes[15,16].

This result bears a direct proportionality to the polarization of the empty states and opens perspectives of quantitative measurements of magnetic effects in empty states, and of Auger weight in the total yield.

The technique couples all the advantages of XAS in terms of dipole selection of final states, sensitivity to small amounts of atoms in a matrix or at a surface, sensitivity to details of energy distribution of empty DOS, and the power of spin resolution for analysing the magnetization of solids. The total yield technique furthermore is not limited to surface systems but can allow the study of subsurface systems, i.e magnetic samples covered by non magnetic or magnetic adlayers.

Acknowledgements:

We sincerely thank Martin Landolt who made available the Mott detector for the preliminary set-up described in this paper and Hans C. Siegmann for overall support. This work was partially supported from the Schweizerischer Nationalfonds under contract NPF 24 and for one of uf (F.S.) by Italian National Research Council (CNR).

References

1) O.Paul, M.Taborelli, M.Landolt,Surf. Sci. **211/212**, 724 (1989)

2) B.Sinkovic et al.,Phys.Rev.Lett. **65**, 1647 (1990)

3) H.C.Siegmann, F.Meier, M.Erbudak, and M.Landolt;Advances in Electronic and Electron Physics, **62**, 1 (1986)

4) LURE 1985-1987 activity report CNRS MEN CEA, p.69

5) S.W.Kortboyer, J.B.Goedkoop, F.M.F.De Groot, M.Grioni, J.C.Fuggle, H.Petersen, Nucl. Instr. and Meth. **A275**, 94 (1989).

6) Electron Energy Analyser MAC II, Riber.

7) 775 Programmable Counter/timer, Keithley.

8) SR530 Lock-in amplifier, Stanford Research System.

9) N.F.Mott, Proc. R. Soc. London Ser. A **124**, 425 (1929)

10) Camberra preamplifier Model 2003BT

11) 994 Dual Counters/Timer, Ortec

12) 37204 HP-IB extender, Hewlett Packard

13) D.T.Pierce and H.C.Siegmann; Phys.Rev. B **9**,4035 (1974)

14) O.M. Paul, PhD thesis ETH Zurich, 1990

15) H.Mizuta, A.Kotani; Journal of the Phys. Soc. of Japan, **54**,4452 (1985); Solid State Commun. **51**, 727 (1984)

16) R.Allenspach, D.Mauri, M.Taborelli, and M.Landolt; Phys. Rev. B **35**, 4801 (1987)

OBSERVATION OF DICHROISM
IN THE 3d-4f TRANSITION
OF HOLMIUM ON SILICON (111)

O. Sakho, M. Sacchi, X. Jin*, F. Sirotti, G. Rossi**

Laboratoire pour l'Utilisation du Rayonnement Electromagnetique CNRS, CEA et MENJ, Université de Paris Sud, F-91405 Orsay

*and Physics Departement Fudan University, Sanghai, China

**and Laboratorium für Festkörperphysik, ETH Zürich, CH-8093 Zürich

Abstract

Linearly polarized light is used as excitation source to study the transition 3d-4f of Holmium on Silicon (111). A strong dichroism is observed, which depends upon thermal treatment and coverages.

Introduction

The dichroism is the dependence of the absorption spectrum upon the polarization of the light. It was first discovered by Biot in 1815 by observing a change of color in tourmaline minerals when the polarization of the light is varied (1).

The origin of dichroism is a splitting of the ground state due to either a crystal field effect, a surface effect, or a Zeeman effect when a magnetic field is present. The dipole transition selection rules acting on the unequal population of the split ground state determine a dichroic absorption in the case of linear or circular polarized light for thermal energies small with respect to the splitting.

Recently, calculations of dichroism in XAS have been stimulated by the novel availability of intense polarized X-rays and soft X-rays from sychrotron sources (2). In particular the atomic like transitions between levels 3d and 4f in rare earths have been calculated for a Zeeman split ground state by Thole et al (3).

The results show that the dipole selected final state multiplets are well separated in the spectra obtained with the polarization of the light parallel or perpendicular to the external magnetic field applied to the rare earth ion. Therefore XAS is very sensitive to ground state splitting.

Experimental confirmation of magnetic dichroism in the 3d-4f transitions of Tb in Tb-Fe garnets was obtained by van der Laan et al (4).

Recently XAS dichroism has been applied to surface mangetic systems: Dy/Ni(110); Tb/Fe(110); Tb/Ni(110) (5,6).

In this case, the surface of a saturated ferromagnetic substrate in which the chemisorbed rare earth ions lie, provided the magnetic field.

We present here the observation of strong dichroism in 3d-4f transitions of Ho/Si(111) where the substrate is not magnetic. This imposes to consider other effects to explain the splitting of the ground state of Ho. We evaluate the role of the crystal field effect.

Experiment

Linear polarized soft X-rays in the energy range 800-1500 eV were available at the outlet of an ultra-high vacuum double crystal monochromator equipped with beryl ($10\overline{1}0$) crystals on a bending magnet of the SuperAco storage ring at LURE (Orsay).

Sample preparation and measurements were performed in a UHV surface science chamber pumped by ionic and turbomolecular pumps.

A manipulator allowed the surface ot the sample to be aligned almost parallel or perpendicular to the X-ray polarization vector. The angle a between the normal to the surface and the polarization of the light was varied between 10 to 90°.

The absorption was measured in the total electron yield mode by using a channeltron, working in the pulse counting mode.

n-type Si(111) single crystal wafers with 5N purity were used as substrates, and could be Ar-ion sputtered and annealed by direct ohmic heating. 4N purity Ho samples were evaporated from thoroughly outgassed tungsten baskets. The evaporation rates were monitored by a quarz microbalance. The annealing temperatures were measured by an optical pyrometer.

Results and comments

In this section we present at first, a description ot the Ho multiplet and then we present our results and try to explain them.

The M_5 spectrum of Ho displays three main structures (7), well separated in energy, which mainly correspond to transitions selected by the dipolar selection rule on J (Fig. 1).

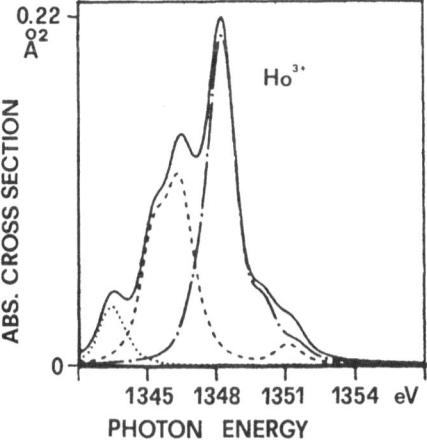

Fig. 1
Decompsition of Ho^{3+} in 3 contributions depending on the dipolar selection rules on J: the component at the left for $\Delta J=1$; in the middle for $\Delta J=0$; at the right for $\Delta J=-1$. According to ref. (7).

In presence of a quantization axis, the I_8 ground state is split into sublevels characterized by the quantum number M. The selection rule on M depends upon the polarization of the light. For linearly polarized light, transitions with $\Delta M=0$, or +/-1 are allowed if the polarization of the light is respectively parallel or perpendicular to the quantization axis, conversely for resolving $\Delta M=+/-1$ transitions, circular polarization is necessary,

In our case, we work with linearly polarized light. Based on cross section calculations taking into account the selection rule on J and M (1), we expect to observe transitions depending on the polarization of the light in the part of the spectrum which satisfies $\Delta J=0$, $\Delta M=0$ if the ground state splitting is due to a magnetic field oriented in the surface plane. Actually it can be shown that a crystal field with a strong axial symmetry oriented perpendicular to the surface would induce qualitatively the same dichroism in the spectra (8).

Our experiment consists in measuring the M_5-edge absorption as a function of the angle α for various thermal treatments of the Ho/Si(111) interfaces.

Submonolayers of Ho (0.3 Å) deposited on Si(111) at room temperature (Fig. 2) show large dichroism, which increases further with annealing (Fig. 3).

For 30 Å of Ho deposited on Si(111) at room temperature (Fig. 4), we observe a reduced dichroism with respect to the submonolayer. Annealing at 600 °C increases the dicroism similarly to what has been found for Dy/Si(111) interfaces (9).

All the spectra show higher integrated intensity at $\alpha=90°$ than at 10°. The main spectral change is for the part of the spectrum corresponding to transitions with $\Delta J=0$.

The cross section calculations for each allowed ΔJ as a function of ΔM for a ground state with M=0 (1) show that the largest dichroism is expected for $\Delta J=0$ transitions, while the $\Delta J=+/-1$ transitions are almost polarization independent.

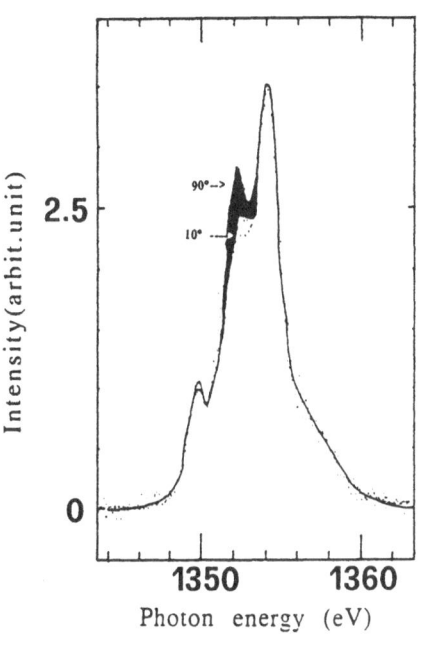

Fig. 2
3 Å Ho/Si(111) as deposited at room
temperature

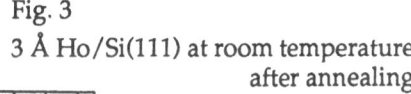

Fig. 3
3 Å Ho/Si(111) at room temperature
after annealing

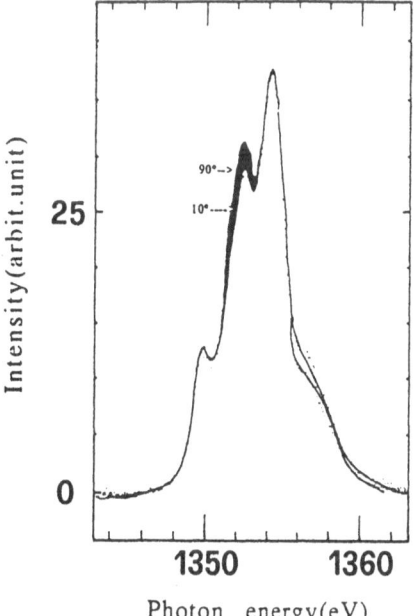

Fig. 4
30 Å Ho/Si(111) as deposited at room temperature

We can empirically separate the multiplets undergoing dichroism in our experiment by substracting spectra obtained with different α from the unpolarized
(non-dichroic) spectrum of bulk Ho. The results are shown in Fig. 5, and compared with the calculated dichroism for Zeeman split Ho in a magnetic field (Fig.
1). Although no external magnetic fields are present in our experiment, we see a
correspondance between the derived peaks and the calculation.

Ho grows epitaxially on Si(111) by nucleating as a compound of stoichiometry
$HoSi_{2-\delta}$ has a hexagonal structure with alternating sixfold coordinated silicon
and holmium planes. Antiferromagnetic ordering of $HoSi_{2-\delta}$ is known with
$T_N=18K$ (12).

The strength of the dichroism at Ho/Si(111) interfaces correlates with the degree
of interface reaction and ordering. In fact no annealed deposits of 30 Å show a
reduced dichroism whilst the largest effect is found for the annealed submonolayer. This would signify that 2-dimensional ordering at the surface or in
the structure of Ho-silicide is important to determine the local field responsible
for the ground state splitting. A large axial component is present in the local
symmetry for surface Ho atoms or Ho sites in epitaxial $HoSi_{2-\delta}$. This is likely to
be the origin of the dichroism. A theoretical effort must be undertaken in order
to evaluate the crystal field effects in geometries relevant to surface/interface
studies, since the sensitivity of XAS to the local field is proven to be very large.

Conclusion

This experiment has shown that it is possible to have dichroism on a non
magnetic substrate and this dichroism is connected to local 2-dimensional order
at the interface.

The sensitivity in XAS of dichroism to local fields at rare earth sites makes this
technique appealing for studying absorbates and interfaces.

With a better quantitative understanding of crystal field dichroism, XAS studies

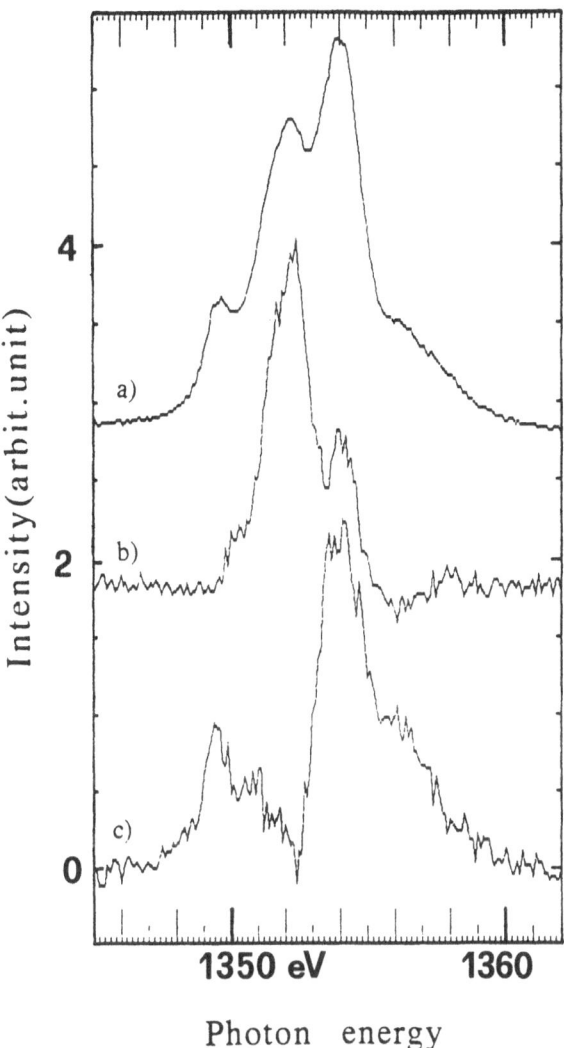

Fig. 5

a) M5 unpolarized spectrum of Holmium; b) part of the spectrum with the polarization of the light parallel to the quantization axis; c) with the polarization perpendicular to the quantization axis

as function of coverage and annealing could become a very powerful tool to probe the degree of stoichiometric and structural order at rare earth/silicon interfaces.

References

(1)   J.Goedkoop, PhD thesis, University of Nijmegen 1989.
(2)   K.J. Kim, Nucl. Instr. Meth. 219 425 (1984).
(3)   B.T. Thole et al. Phys. Rev. Lett. 55 2086 (1985).
(4)   G. van der Laan et al. Phys. Rev. B34 6529 (1986).
(5)   R. Kappert et al. to be published in Surface Science.
(6)   J. Goedkoop et al. submitted to Phys. Rev. B.
(7)   J. Goedkoop et al. Phys. Rev. B37 2086 (1988).
(8)   M. Sacchi et al. to be published in Surface Science.
(9)   M. Sacchi et al. submitted to Phys Rev. B.
(10)   V.M. Kolesho et al. Thin Films 141 277 (1980).
(11)   J. Knapp et al. Materials Research Society Pittsburgh PA vol 54 (1986).
(12)   J. Pierre et al. Journal of the Less Common Metals 139 321 (1988).

Electronic Structure

INVESTIGATIONS ON THE BANDSTRUCTURE
OF
SEMIMAGNETIC SEMICONDUCTORS

H.-U. Middelmann and H.-E. Gumlich

Institut für Festkörperphysik der Technischen Universität Berlin

Hardenbergstraße 36, 1000 Berlin 12, Germany

INTRODUCTION

After the III-V-compound semiconductors e.g. GaAs with their large field of application in electronics and optoelectronics /1/, now the II-VI-compound semiconductors gain increasing interest. The most interesting materials in this field are the semimagnetic semiconductors (SMSC) or diluted magnetic systems (DMS) /2/ which means II-VI-compounds containing large amounts of magnetic ions, like iron or manganese in e.g. $Zn_{1-x}Mn_xSe$. These materials show variable band gaps and additional magnetic effects as a giant Faraday rotation. They are promising candidates for constructing novel magnetically controlled optoelectronic devices. These new effects in the SMSCs are ascribed to the interaction between the manganese 3d electron states with the anion p states of the diamagnetic host crystal. Semimagnetic semiconductors exhibit characteristic properties of luminescence due to the interaction between the Mn^{2+}-ions and the donors or acceptors in the crystal as well as due to transitions within the Mn d states /3/.

The electronic properties of solids are given by the specific bandstructures, which are derived from s- and p-states of the constituent atoms, in the case of II-VI-compounds. The cation d-levels (Zn 3d, Cd 4d) with binding energies in the order of 10 eV show no or nearly no dispersion, i.e. no contribution to the valence band. In the SMSCs where the cations of the II-VI-compounds are partly substituted by manganese the question arises to what extend the paramagnetic Mn 3d states contribute to the host valence band.

With the means of Synchrotron radiation, the reach of experimental methods is drastically extended. Certain experiments, as resonant photoemission or the determination of bandstructure by normal electron emission are feasible, since this tuneable light source with its large range of energy is available. In our group, we use these new possibilities to investigate the bandstructure of SMSC by the method of reflection spectroscopy at normal incidence and by photoemission spectroscopy. Our experiments are performed at the Berlin storage ring BESSY using monochromators within the ranges between 4 and 25 eV (3m-NIM-1), 10 and 120 eV (TGM-1), and 300 and 700 eV /4/.

REFLECTION SPECTROSCOPY

The advantage of reflection spectroscopy is, that especially the fundamental gap and transitions to final states below the vacuum level can be determined, as well as critical points in the Brillouin zone in general and excitonic transitions. Corresponding experiments have been performed in our group on CdTe, $Cd_{1-x}Mn_xTe$, and other SMSCs /5/.

The experimental set up in its main components conceived to that end /6/ consists essentially of the UHV-chamber containing the sample manipulator and a rotating mirror, which is controlled by the computer (fig. 1). The rotating mirror enables us to detect quasi simultaneously the intensities of both, the incoming light $I_i = I_0(\omega)*f(\omega)$ and the reflected light $I_r = I_R(\omega)*f(\omega)$. Here $f(\omega)$ represents unknown influences on the measurement. Higher orders of diffraction due to the monochromator are suppressed by appropriate filters. The signal is detected by a photomultiplier outside the chamber via a Sodium salycilate layer which reduces the photon energy to be transmitted through the window. The intensities I_r and I_i are registrated by the computer, which forms the absolute reflectance R according to $R(\omega) = I_r / I_i = I_R/I_0$.

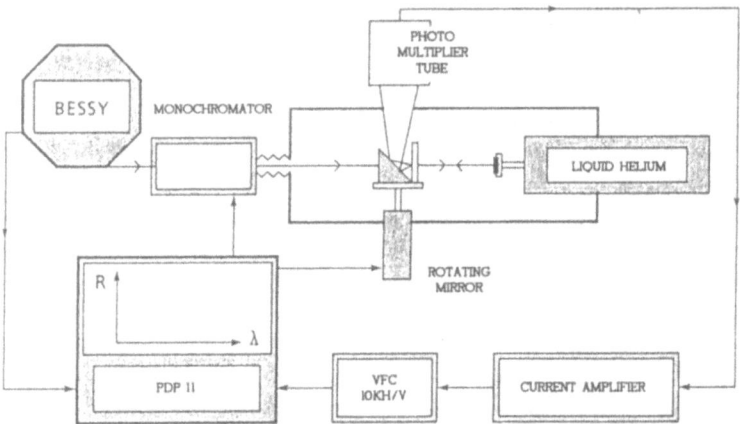

Fig. 1   Reflection spectroscopy, schematic diagram of the experimental set up.

Information on the band structure is achieved by using the aquivalence between the imaginary part of the dielectric function $\varepsilon_2(\omega)$ and the joint density of states (J_{VC}) :

$$\varepsilon_2 = \frac{e^2}{\pi m^2 \omega^2} \int |a_0 \cdot p_{ji}|^2 \frac{dS}{|\nabla_k E_{ji}(k)|} \qquad J_{VC} = \frac{1}{8\pi^3} \int \frac{dS}{|\nabla_k E_{ji}|_{E_{ji}=h\nu}}$$

where j and i denominate unoccupied and occupied states, respectively and dS is a surface element in k-space such that $E_{ji}(k) = h\nu$ /7/.

Spectral features, correlated with points of high density of states, show characteristic line shapes corresponding to the van Hove singularities in the joint density of
states. The type of transition is analysed by a suitable line shape fit from which the
type of critical point in the BZ is derived /8/.

The dielectric function $\varepsilon(\omega)$, which is relevant for the determination of bandstructure
data from experimental reflectivity, and the other optical functions e.g. index of refraction $n(\omega)$ or reflectivity $R(\omega)$ are related by anlytical formulas, i.e. if one of these
functions is determined, all the others are to be calculated /9/.

In such a manner from the known complex reflectivity $r(\omega)$ real and imaginary part
of the dielectric function are derived by

$$\varepsilon_1(\omega)= \frac{(1-R)^2-(2\sqrt{R}\sin\Theta)^2}{(1+R-2\sqrt{R}\cos\Theta)^2} \qquad \varepsilon_2(\omega)= \frac{(1-R)(2\sqrt{R}\sin\Theta)}{(1+R-2\sqrt{R}\cos\Theta)^2} \qquad (1)$$

where $R = R(\omega)$ is the experimentally obtained absolute reflectivity and $\Theta = \Theta(\omega)$ is the
phase of the complex $r(\omega) = \sqrt{R}\,e^{i\Theta}$. The calculation of ε from R obviously requires
further information, which is not given by the experiment. While from ellipsometric
measurements both real and imaginary part of the complex optical function are
derived directly from the experiment, reflection measurements in normal incidence
deliver only the absolute reflectivity but not the phase $\Theta(\omega)$ of the complex reflectivity.
Nevertheless the complex optical functions can be obtained for instance by applying
the Kramers Kronig relations /10/ :

$$\Theta(\omega)= -\frac{\omega}{2\pi}P\int_0^\infty \frac{\ln R(\omega')}{\omega'^2-\omega^2}d\omega' \qquad (2)$$

Transformation of the Cauchy principal value integral results in

$$\Theta(\omega)= -\frac{1}{2\pi}\int_0^\infty \left(\frac{d}{d\omega'}\ln\left|R(\omega')\right|\right)\left(\ln\left|\frac{\omega'+\omega}{\omega'-\omega}\right|\right)d\omega' \quad . \qquad (3)$$

Data of R for energies between zero and infinity are nessecary for this integration.
That means, the integration has to be performed by using an appropriate extrapolation
of $R(\omega)$ for energy values which exceed the experimental range.

Different methods are used to find a suitable "tail function". Our method, which
results in a correct reproduction of the dielectric function in the whole experimental
range is based on an iterative procedure in addition to the commonly used approximation derived from Drude's theory:

$$R(\omega); (\omega \geq \omega_2) = R(\omega_2)*h\omega_2\big/(h\omega)^x,$$

where x represents a real number. Sometimes x is used as a fitparameter. However,
provided that no further oscillators beyond ω_2 take effect, after Drude's theory this
exponent should be $x=4$ /11/.

With the chosen tail function of R for $\omega \geq \omega_2$ and its definition in the low energy
region from $\omega=0$ to ω_1 either by data taken from other experiments or by approximation as a linear function, the phase function $\Theta(\omega)$ can be determined via the

Kramers Kronig transformation, i.e. by integration of equation (3). Then from R and Θ the dielectric function $\varepsilon(\omega)$ is to be calculated after equation (1).

In a next step the dielectric function, obtained in this manner, is compared with the absolute reflectivity obtained by the experiments. To that end the imaginary part ε_2 is approximated by a great number (in the order of $n=50$) of Lorentz oscillators in such a way that deviations between the KKR-derived function and the approximated function are minimized, keeping in mind the experimental error. In order to realize this large scale fit within a satisfactory time, a computer program was developed, the algorithm of which relies on Darwin's principle of evolution /12/.

After this procedure ε_2 is reproduced as a sum of n Lorentzians with a frequency ω_k, an oscillator strength f_k and a full width of half maximum Γ_k. Since in the complex Lorentz function both the imaginary part ε_2 and the real part ε_1 are expressed by these parameters /13/, now the complete complex dielectric function is reproduced:

$$\varepsilon_{Lorentz} \sim \sum_{k}^{n} \left\{ f_k \frac{\omega_k^2 - \omega^2}{(\omega_k^2 - \omega^2)^2 + \omega^2 \Gamma_k^2} + f_k \frac{i \Gamma_k \omega}{(\omega_k^2 - \omega^2)^2 + \omega^2 \Gamma_k^2} \right\} \quad (4)$$

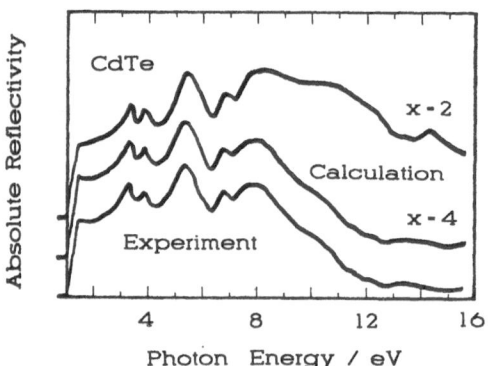

Fig. 2 Experimental and calculated absolute reflectivity showing the influence of different extrapolations used in Kramers Kronig analysis.

Using the dependence $R=R(\varepsilon)$ the corresponding absolute reflectivity is calculated, which is compared with the experimental reflectivity (fig. 2). No deviation should occur especially in the high energy region if a correct tail function was used. As is shown by our results for CdTe, there are significant deviations in the range above 6 eV for the exponent $x=2$ but a good overall agreement between experiment and reproduction for $x=4$. We think that this method delivers a good criterion for the correctly chosen extrapolation and the quality of the dielectric function derived via Kramers Kronig transformation. A complete description of this investigation on CdTe and the assignments of the structures in the reflectivity spectra is given in ref. /14/.

PHOTOEMISSION

Since the synchrotron radiation is increasingly available the angle resolved photo-electron spectroscopy (ARPES) /15/ represents by now a well established tool for the experimental determination of the electronic bandstructure $E(k)$ of semiconductors /16/. With excitation energies ranging from 10 to 120 eV, new bandmapping techniques are feasible. Before this tuneable light source was available, in the far vacuum ultra-violet (VUV) regime above 11 eV only discrete sources, like He-resonance lamps, could been used. Experiments with such sources requires measurements of electron energy distribu-tion curves taken at different angles of emission, in order to obtain reliable informa-tion on $E(k)$-dispersion. A correct assignment of the k-vectors /17/ is comparatively complicated, because of the deflection of electrons crossing the surface barrier, which is caused by the different potentials inside and outside the crystal. As a result the momentum of the photoemitted electron is changed, i.e. the component normal to the surface k_s is reduced while the parallel component k_p remains constant. That means in general there are different directions of k inside and outside the crystal.

This problem is eliminated by using a tuneable lightsource. By detecting electrons only emitted normal to the surface, the direction of k is well defined by the orien-tation of the crystal, since k_p = const. = 0. Instead of the emission angle now the photon energy is varied, whereby k-vectors are obtained which cover the whole Brillo-uin zone in the direction of detection.

Such experiments concerning compound semiconductors have been performed first on III-V- compound semiconductors e.g. GaAs, GaP, InSb, InP /18/ . In the meantime bandmapping experiments using normal emission are performed on II-VI- compound semiconductors as well. Chab and coworkers /19/ investigated the bandstructure of CdTe(110) along the ΓKX -(Σ) direction of the BZ and found a good overall agreement with the relevant calculations. At the same time they checked the influence of Manga-nese on the bandstructure of $Cd_{1-x}Mn_xTe$. Beside a broadening of structures in general and an additional contribution at a binding energy near 3.5 eV they found no significi-ant change of the dispersion $E(k)$ compared to CdTe.

In order to investigate the contribution of manganese states to the valence band, the method of resonant photoemission /20/ is advantageous, due to the selective enhance-ment of manganese 3d- derived photoemission. From states, strongly localized at the specific atoms (i.e. manganese 3p-core levels), transitions are excited to unoccupied Mn d-states near the Fermi level, which are more or less hybridized with the outer p-sta-tes of the surrounding anions. Depending on the degree of hybridization, different strengths of resonance are obtained. The photon energy required for exciting the mangnese 3p - 3d transition has a value near 50 eV. This intermediate excited state decays by auto-ionization, exciting a d-electron into continuum states well above the vacuum level of the crystal:

$$3p^6 3d^5 + h\nu \rightarrow 3p^5 3d^6 \rightarrow 3p^6 3d^4 + e.$$

Since in addition a likelihood is given for the direct excitation of a d-electron by the incident photon,

$$3p^6 3d^5 + h\nu \longrightarrow 3p^6 3d^4 + e$$

with both processes having the same initial and final system state, a quantum mechanical interference between these two channels occurs. This interaction results in the so called "Fano resonance", which is characterized beside the asymmetric line shape by a resonance maximum and a resonance minimum. The analytic expression of the energy dependence of this type of resonance was derived by Fano /21/, corresponding to

$$I(h\nu) = I_0 (h\nu) \frac{(E+q)^2}{E+q^2} \qquad \text{with} \qquad E = \frac{h\nu - E_0}{\Gamma/2}$$

The asymmetry parameter q involves the ratio of the contribution of the two excitation channels participating at the resonance, where q increases with increasing auto-ionization. The behaviour of the Fano curve and its extrema dependent on q is shown in fig. 3. For increasing q the position of the maximum converges at E_0 while the minimum position shifts to lower energies. In the extreme case with q going to infinity, the resulting line shape is represented by a Lorentz curve with the maximum E_{max} at E_0 and the full width of half maximum Γ.

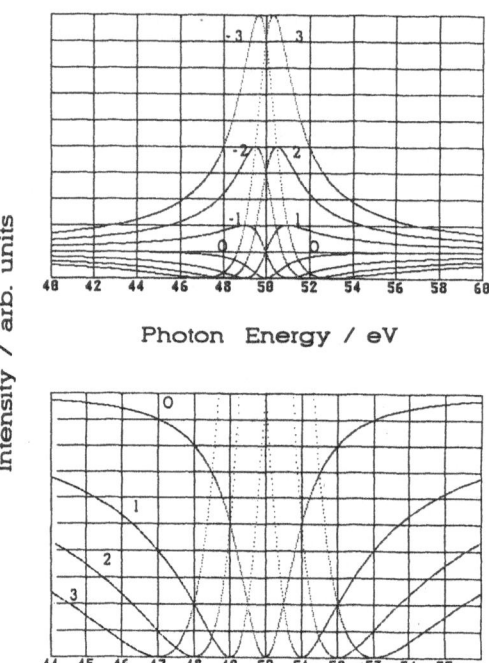

Fig. 3   Calculated Fano line shapes.
Dependence on the
asymmetry parameter q
for $I_0 = 1$, $\Gamma = 2$ eV,
and $E_0 = 50$ eV.

A survey concerning problems due to manganese d states and their investigation using photo electron spectroscopy in different SMSCs was given by Wall et al. /22/.

For $Cd_{1-x}Mn_xTe$ exemplary resonant photoemission experiments were performed by Ley and Taniguchi et al. /23/ proceeded by corresponding investigations on $Cd_{1-x}Mn_xSe$ and $Cd_{1-x}Mn_xS$ /24/. The resonant spectra were interpreted in terms of configuration interaction (CI) calculations by Fujimori et al. /25/, which appear to supply convincing assignments of the observed features.

For our resonant measurements on $Zn_{1-x}Mn_xY$ (Y=S,Se,Te) semimagnetic semi-conductors we used bulk crystals and for comparison thin films, grown ex situ epitaxi-cally on ZnSe (111) by electron beam evaporation. The bulk samples were cleaved prior to insertion into the vacuum chamber. The basic pressure was in the lower region of 10^{-10} mb. All samples were cleaned in situ by weak bombardment with Ar ions (500 eV) followed by an annealing at temperatures near 500 K. Since our intention was not to do bandmapping experiments but rather experiments, based principally on localized effects, the orientation of the single crystal domains was less relevant. The aim of our experiments was to detect changes in the density of states due to the contribution of Manganese. The set-up of our experiments is shown in fig. 4.

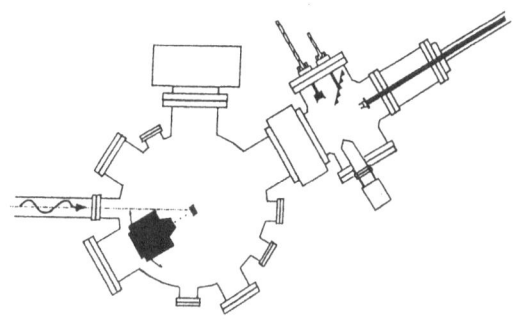

Fig. 4   Photoemission,
schematic diagram
of the experimental set up.
upper part:
Ultra high vacuum
electron spectrometer
with preparation chamber.
lower part:
Hemispherical
electron energy analyser
with data aquisition

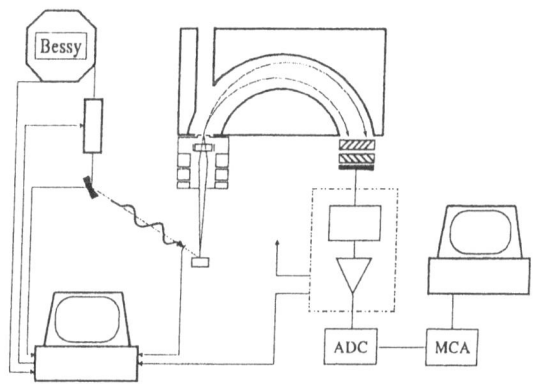

A typical series of energy distribution curves for different photon energies near the resonant Mn 3p-3d excitation is shown in fig. 5. The spectra were taken from $Zn_{1-x}Mn_xSe/ZnSe(111)$ thin films /26/. The dominant feature at a binding energy of 9.5 eV is due to photoemission from the Zn 3d (d-final state) core level. Its intensity is nearly constant in the whole range of photon energy. The maximum at 3.6 eV de-pends strongly on the excitation energy. A plot of the amplitude of this peak, after background substraction, versus the photon energy delivers the typical Fano line shape

with a maximum near 50 eV and a minimum near 48 eV (fig. 6, upper part). To
eliminate intensity changes due to variing monochromator transmission the emission
was normalized to the Zn 3d - emission, the intensity of which is shown to be nearly
constant in the considered region /27/,/30/.

A comparison of the Mn 3p - 3d resonance from $Zn_{1-x}Mn_xSe$, obtained in the
described manner, with the resonant behaviour of atomic manganese /28/ delivers
a nearly identical line shape of both curves (fig. 6, lower part). This can be interpreted in that way, that there is no hybridization of those d-states, involved in this analysis i.e. at the binding energy of 3.6 eV.

Fig. 5   Fig. 6

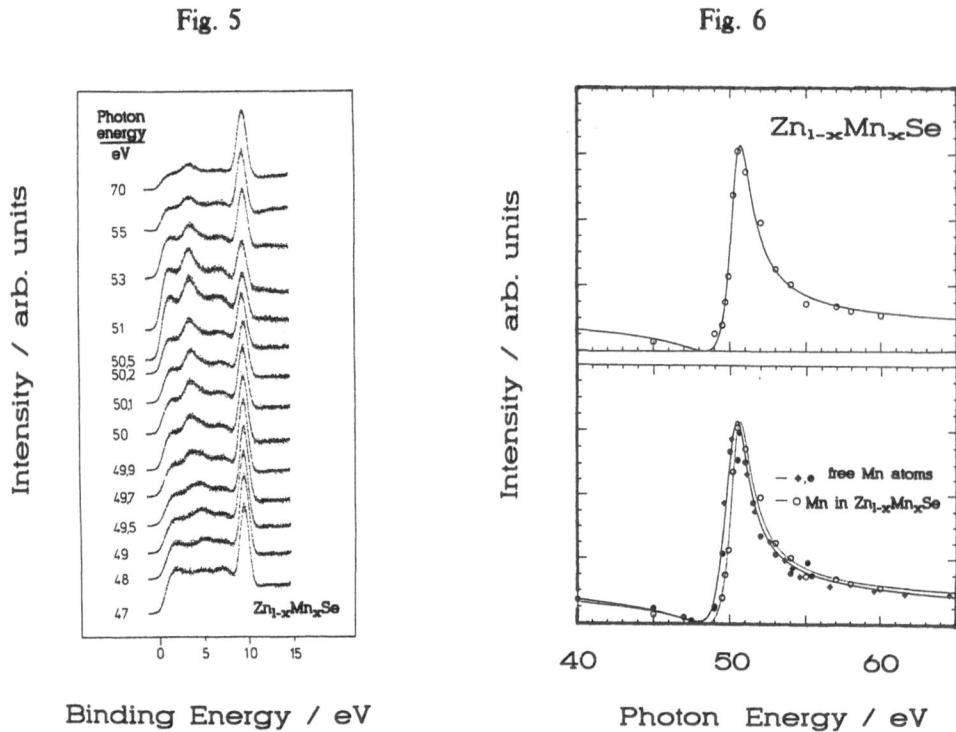

Binding Energy / eV   Photon   Energy / eV

Fig. 5   Series of electron energy distribution curves for different photon energies,
          showing the valence band region at the Mn 3p-3d resonance.
Fig. 6   upper part: The resonant peak at 3.6 eV (see Fig. 5).
                      Dependence on excitation energy and a Fano line shape fit.
          lower part: Comparison with data obtained from atomic manganese
                      from Kobrin et al. /27/.

This finding supports some of the results from ligand field theory which predicts for
zinc blende structure crystal field split d- states with the two different symmetries E,
and T_2, from which the E-states should be localized and the T_2-states should be mixed

with anion p-states /29/. Accordingly further Mn 3d contributions to the valence band are to be expected. In order to find out such contributions in the EDCs it is useful to form difference spectra. Since the resonant photoemission delivers both a large enhancement of Mn 3d derived emission and a deletion due to the quantum mechanical interference as well, a difference spectrum between EDCs taken at resonance maximum and resonance minimum should reflect only structures caused by manganese (fig. 7). In principle the other structures in the valence band region can change as well. This effect would falsify the results. However, in so far as changes are considered which are caused by bandstructure effects, i.e. peak shifts or intensity changes, they can be neglected because of the comparatively large ratio of the used excitation energies (near 50 eV) and the difference energy (2 eV).

The strongest feature in the difference spectrum is the maximum at 3.6 eV. Additional structures can be identified at 1 eV and near 7 eV, which might represent p-d-hybridized portions with T_2-symmetry according to ligand field theory. A hint for correct interpretation could be given by analysing the resonant behaviour of these additional structures. However, because of the bad signal to noise ratio in the region out of the resonance maximum this is more difficult as it is for the main peak at 3.6 eV.

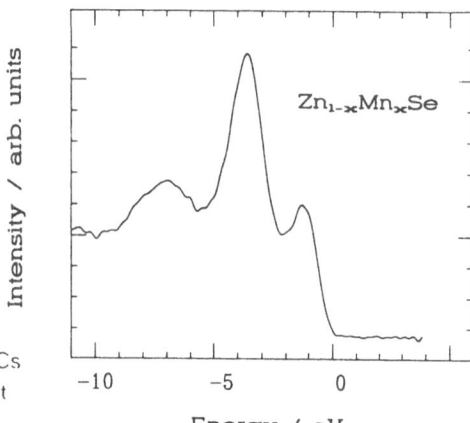

Fig. 7   Difference spectrum obtained by EDCs
          taken at resonance maximum and at
          resonance minimum,
          showing the Mn 3d contribution.

An elegant possibillity is given by the method of constant initial state spectroscopy (CIS) where the emission of a selected constant initial state for variing excitation energies is detected by synchronously changing of the photon energy with the analyzer energy keeping the difference of both energies constant: $h\nu - E_{kin}$ = const. Unfortunately this method implies of course also the detection of the background, which may cause significant deviations in the case of a nonmonotonous background. At least for a CIS spectrum taken in the region near 7 eV, i.e. at kinetic energies below that of the other resonant structures, one has to expect a resonant background which corresponds essentially with the main peak emission and exhibits a similar resonance. An alternative, which needs some more effort, would be given by taking spectra of $Zn_{1-x}Mn_xSe$ as well as of ZnSe at various photon energies in the region of interest and by subsequent formation of a series of difference spectra from EDCs obtained at the same photon energy in each case. These difference spectra can be used to derive photon energy dependencies at specific binding energies, which can be analyzed in terms of Fano line shape. Here of course other shortcomings are to be taken into account, but as we see from our latest results obtained on different SMSCs /30/, this type of difference spectra at least delivers comparable results for photon energies at 50 eV (resonance maximum).

However, the broad elavation near 7 eV could not be explained in the one electron picture by bandstructure calculations /31/, as was found for $Cd_{1-x}Mn_xTe$ /23/. But it appears to be sure that the structure near the valence band maximum at 1 eV is caused by p-d hybridization. Due to the possible screening by charge transfer from the anions to the Mn d states, the photo electron then reflects a d^5-final state. Thus a certain comparability with results obtained by luminescence seems to be justified, which are assigned to Maganese d- levels with binding energies near 1eV as well /32/.

The strength of screening at the photoemission process consequently depends on the degree of hybridization of anion p- and Mn d-states (T_2 and E).

According to ligand field theory thus a transfer to d-states with T_2-symmetry is expected. Accordingly the resulting structures should represent d^5-final states and the structures near 7 eV should reflect unscreened states with T_2 symmetry. By the same symmetry considerations contributions from E-states would represent d^4-final states. With increasing hybridization an enhanced screening is given so that in reference to d^4-related emissions d^5-related ones increase. At the same time the resonant behaviour, which is favoured by the localized states, decreases.

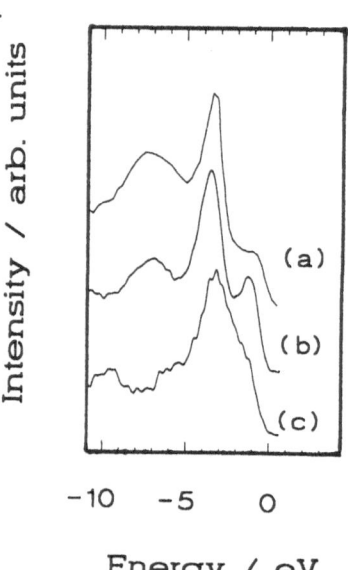

Fig. 8   Difference spectra,
          showing the Mn 3d energy distribution
          dependent on the anion
          (a) $Cd_{1-x}Mn_xTe$, from Ley et al./22/,
          (b) $Zn_{1-x}Mn_xSe$,
          (c) $Zn_{1-x}Mn_xS$.
          Decreasing photoemission is detected
          going from Te to S.

This tendency is to be observed in our difference spectra for $Zn_{1-x}Mn_xSe$ and $Zn_{1-x}Mn_xS$ /33/. Corresponding spectra are shown in fig. 8. For comparison we use the difference spectrum of $Cd_{1-x}Mn_xTe$ from Ley et al. /23/. With the expected increasing p-d- hybridization going from Te to S, in the difference spectra a decreasing tendency for the broad contribution around 7 eV is detected. This behaviour is explained in that way, that these emissions represent unscreened states /34/. The interpretation of this finding as a consequence of different degrees of hybridization is consistent with the behaviour of the q-parameters of the corresponding Fano curves. With increasing hybridization we find decreasing values for q /30/.

Using the configuration interaction theory, the features in the photoemission spectra are explainable within one model /23/, leading to the result, that beside charge transfer to T_2 states also charge transfer to E states occur. That means that both states in the spectra are represented by d^5-final states and thus have the same reference energy. By the same argument the region of unscreened states reflects d^4-final states, concerning to T_2 as well as to E symmetry. In the ligand field model, the E (d^4) states would have a reference level different from the hybridized T_2 (d^5) states due to different total energies corresponding to the different final states /36/.

Beside the resonant excitation of the Mn 3p-3d transition, which delivers assessments on the Mn contribution to the valence band of the considered semimagnetic semiconductor, excitation of the Mn 2p-3d transition gives information on the ground state of the manganese ion. For a sufficiently high resolving power parameters on the effective crystal field are obtainable. This was shown by F.M.F. de Groot during this workshop by calculations and experimental results, see also Ref. /36/.

In order to excite transitions in this region photon energies in the range from 630 to 660 eV are required. We have performed such measurements at the high energy grating monochromator (HE-TGM-2) at BESSY for various SMSCs. As experimental method predominantly the partial yield spectroscopy was used, i.e. the analyzer was fixed at a certain kinetic energy in the region of secondary electron cutoff. This leads to a drastical enhancement of the signal to noise ratio. Dependent on the detected kinetic energy of the secondary electrons one obtains emphasized volume- or surface information. The spectra presented here were taken at low kinetic energies near 1 eV whereby a relatively large volume sensitivity was obtained. From the basic mechanism those measurements are compareable with absorption measurements /37/, but the partial yield method has the advantage that the sample thickness is irrelevant.

In order to review our measurements our results for the different materials are shown within a large energy interval going from 500 to 700 eV in fig.9.
For comparison beside the spectra taken from SMSCs additional measurements on metallic manganese and ZnO were performed. Common to all the spectra, except for ZnO, is the characteristc dublet structure in the region of 630 to 660 eV, which is due to the spin orbit interaction of the Mn $2p^5$ final state. The spin orbit split energy is 11.2 eV. Strong additional structures are observable in the absorption of $Zn_{1-x}Mn_xTe$ which is traceable to corresponding Auger transitions. A further feature which appears in the spectra for $Zn_{1-x}Mn_xSe$ as well as for the manganese metal sample is explained by

oxygen contamination, which becomes evident by comparison with ZnO. There is no significiant influence by this contamination on the low energy resonance at the 3p-3d treshold as it was confirmed by following specific investigations /38/.

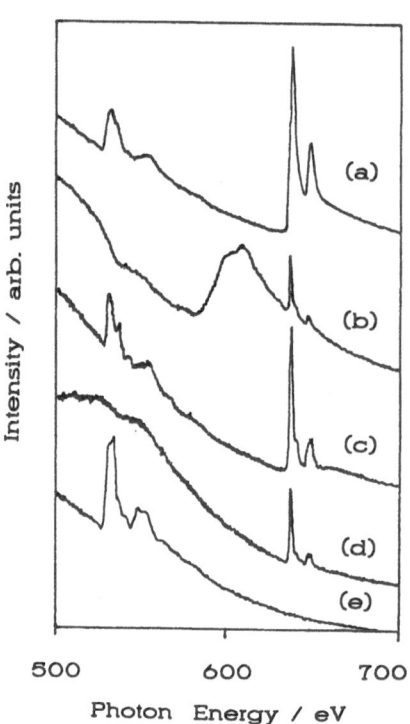

Fig. 9   Partial yield spectra obtained by
detecting secondary photo electrons
for various materials.
Strong structures occur due to
O 1s excitation and Mn 2p excitation.
(a) metallic Mn,
(b) $Zn_{1-x}Mn_xTe$,
(c) $Zn_{1-x}Mn_xSe$
(d) $Zn_{1-x}Mn_xS$,
(e) ZnO.

The region of the Mn 2p threshold for $Zn_{1-x}Mn_xSe$ is shown in Fig 10. Within the two main structures which are assigned to the $p_{1/2}$- and the $p_{3/2}$- state, respectively, additional contributions arise which should depend on both the ground state of the manganese ion in the SMSC and on the effective crystal field /39/.

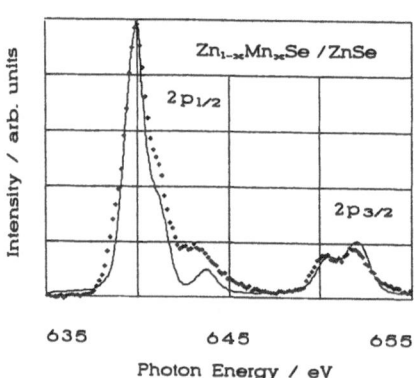

Fig. 10   Secondary yield spectrum
showing
the Mn 2p - 3d excitation
compared with
calculations performed
by Thole et al. /37/
(see also fig. 11).

Electron spectra of manganese without crystal field electron spectra for various possible ground states have been calculated by Thole and coworkers /39/, which are depicted in fig. 11.

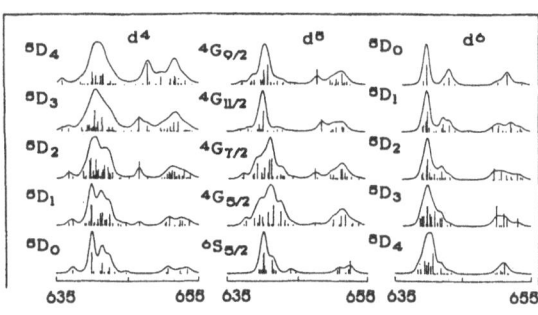

Fig. 11   Electron spectra, calculated for various ground states of atomic manganese, from Thole et al. /37/

Photon Energy / eV

A comparison of our experimental results with this choice of various ground state calculations reveals a good agreement with the calculated spectrum of the $^6S_{5/2}$- ground state for all the SMSCs under investigation. Little differences are observed between the various samples. For a better comparability, together with the calculated curve of Thole et al. the two main regions are shown separately but in the same scale after a linear background subtraction (fig. 12).

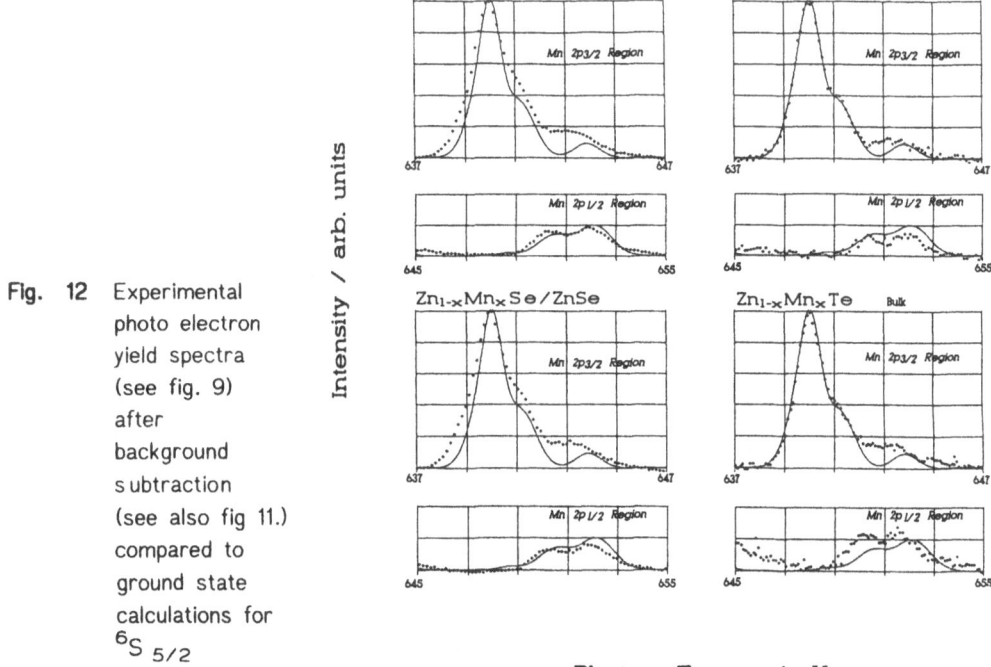

Fig. 12   Experimental photo electron yield spectra (see fig. 9) after background subtraction (see also fig 11.) compared to ground state calculations for $^6S_{5/2}$

Photon Energy / eV

The intensities are normalized to the dominating maximum of the $2p_{3/2}$-region. On the left hand side the results are shown obtained from the thin film crystals $Zn_{1-x}Mn_xS/CaF_2$ with NaCl-structure and $Zn_{1-x}Mn_xSe/ZnSe$ with zinc blende-structure. Between these both materials no significiant differences are detectable. In comparison with the calculation a broadening is noticed, which is not due to the limited resolution of about 1.8 eV, but rather to the lower degree of crystallin order of these samples caused by the method of growing. Obviously that is why the compound specific differences are not resolved. This becomes evident by comparison with the results obtained from the bulk crystals shown on the right hand side of fig. 12. The dominant structure in the $p_{3/2}$-regime demonstrates nearly exact agreement with the calculation. Deviations occur for the following weaker structure. In this range also slight differences between $Zn_{1-x}Mn_xS$ (Wurtzit) and $Zn_{1-x}Mn_xTe$ (Zinc-blende) are detectable as well as in the $p_{1/2}$-regime.

SUMMARY

Synchrotron radiation is a very useful tool to investigate the bandstructure of semimagnetic semiconductors. Reflectivity measurements have been used to examine the influence of the magnetic ions on the bandstructure. A new way of analyzing data obtained by these measurements has been demonstrated. Resonant photoelectron spectroscopy of the transitions 2p–3d and 3p–3d of Mn^{2+} deliver data of the degree of hybridization and of the contribution of the 3d-states of Mn^{2+} to the valence band structure.

ACKNOWLEDGEMENT

This work was partly supported by the Bundesminister für Forschung und Technologie, contract N° 05 414 CAB 7 and by the Technische Universität Berlin, IFP 4/1. We would like to ackknowledge friendly cooperation with U. Becker, B. Burmester, Ch Jung, Th. Kleemann, A. Knack, Th. Kreitler, M. Kupsch, H.-C. Mertins, and R. Weidemann.

REFERENCES

/1/   L. Esaki, IEEE J. Quantum Electronics QE-22 (1986) 1611
/2/   J.A. Gaj, J. Phys. Soc. Japan A49 (1980) 747
       N.B. Brandt, V.V. Moshchalkov, Advances in Physics 33 (1984) 193
       R.R. Galazka, J. Cryst. Growth 72 (1985) 364, Europhys. News 18 (1987) 90
       W. Gebhardt in "Excited-State Spectroscopy in Solids", p. 111, Bologna 1987
       O. Goede, W. Heimbrodt, Phys. Stat. Sol. (B) 146 (1988) 11
       J.K. Furdyna, J. Appl. Phys. 64 (1988) R29

/3/   C. Benecke, H.-E. Gumlich, in "Diluted Magnetic Semiconductors", M. Jain (Ed.), World Scientific Publisher, London 1991

/4/   BESSY, Annual Report 1984

/5/   M. Krause, H.-E. Gumlich, U. Becker, Phys. Rev. B 37 (1988) 6336
       E. Flach, H.-E. Gumlich, Ch. Jung, M. Krause, Phys. Stat. Sol. (B) 155 (1989) 317
       Ch. Jung, H.-E. Gumlich, A. Knack, H.-C. Mertins, J. Cryst. Growth 101 (1990) 926
       Ch. Jung, Th. Bitzer, N. Dietz, H.-E. Gumlich, H.-U. Middelmann, submitted for publ.

/6/   Ch. Jung, Dissertation, Technische Universität Berlin, 1990, D 83

/7/   F. Wooten, Optical Properties of Solids, 1972

/8/   D.E. Aspnes in Optical Properties of Solids, Vol. 2, M. Balkanski (Ed.), North Holland, Amsterdam, 1980

/9/   see e.g. D.W. Lynch in (and references therein) Handbook on Synchrotron Radiation, Vol 2, G.V. Marr (Ed.), North Holland, Amsterdam, 1987

/10/   B. Velicky, Czech. J. Phys. B 11 (1961) 787
       F. Bassani, M. Altarelli in Handbook on Synchrotron Radiation, Vol 1, E.-E. Koch (Ed.), North Holland, Amsterdam, 1983

/11/   D.L. Greenaway, G. Harbeke in The Science of Solid State, B.R. Pamplin (Ed.), Pergamon Press, Oxford, 1968

/12/   N. Dietz, Diplomabeit, Technische Universität Berlin, 1988
       Th. Bitzer, Diplomabeit, Technische Universität Berlin, 1988

/13/   see e.g. J.M. Ziman, Principles of the Theory of Solids, 2nd Edition 1972, Cambridge University Press, London, 1972

/14/   Ch. Jung, Th. Bitzer, N. Dietz, H.-E. Gumlich, H.-U. Middelmann, submitted for publ.

/15/   N.V. Smith, F. J. Himpsel in Handbook on Synchrotron Radiation, Vol. 1a, E.-E. Koch (Ed.) North Holland, Amsterdam, 1983

/16/   T.C. Chiang, F.J. Himpsel in Landolt-Börnstein, New Series, O. Madelung (Ed. in Chief), Vol 23, Electronic Structure of Solids, Photoemission Spectra and Related Data Sub.Vol. a, A. Goldmann, E.-E. Koch (Eds.)

/17/   F.J. Himpsel, Appl. Optics 19 (1980) 3964

/18/   T.-C. Chiang, J.A. Knapp, M. Aono, D.E. Eastman, Phys. Rev. B 21 (1980) 3513
       F. Solal, G. Jezequel, F. Houzay, A. Barski, R. Pincheaux Sol. St. Commun. 52 (1984) 37
       G.P. Williams, F. Cerrina, G.J. Lapeyre, J.R. Anderson, R.J. Smith, J. Hermanson Phys. Rev. B 34 (1986) 5548
       H.-U. Middelmann, L. Sorba, V. Hinkel, K. Horn, Phys. Rev. B 34 (1986) 957
       L. Sorba, V. Hinkel, H.-U. Middelmann, K. Horn, Phys. Rev. B 36 (1987) 8075

/19/   V. Chab, G. Paolucci, K.C. Prince, M. Surman, A.M. Bradshaw Phys. Rev. B 38 (1988) 12353

/20/   L.C. Davis, J. Appl. Phys. 59 (1986) R25

/21/   U. Fano, Phys. Rev. 124 (1961) 1866

/22/   A. Wall, S. Chang, P. Philip, C. Caprile, A. Franciosi, R. Reifenberger, F. Pool J. Vac Sci. Technol. A 5 (1987) 2051

/23/   L. Ley, M. Taniguchi, J. Ghijsen, and R.L. Johnson, and A. Fujimori, Phys. Rev B 35 (1987) 2839

/24/   M. Taniguchi, M. Fujimori, M. Fujisawa, T. Mori, I. Souma and Y Oka,
       Sol. St. Commun. 62 (1987) 431
/25/   A. Fujimori, F. Minami, S. Sugano, Phys. Rev. B 29 (1984) 5225
/26/   B. Burmester, H.-E. Gumlich, Ch. Jung, A. Krost, H.-U. Middelmann, D. Ricken,
       R. Weidemann, U. Becker, M. Kupsch, Proc. 19th Int. Conf. on the Physics on
       Semicond., p. 1583, W. Zawadzki (Ed), Warsaw 1988
/27/   W.R. Johnson, V. Radojevic, P. Deshmukh, K.T. Cheng, Phys. Rev. A 25 (1982) 337
/28/   P. H. Kobrin, U. Becker, C.M. Truesdale, D.W. Lindle
       J. Electron Spec. Rel. Phen. 34 (1984) 129
/29/   S. Sugano, Y. Tanabe, H. Kamimura: Multiplets of Transition-Metal Ions in Crystals,
       Academic Press, New York and London 1970
/30/   R. Weidemann, H.-E. Gumlich, M. Kupsch, H.-U. Middelmann, U. Becker
       to be published elsewhere
/31/   B. Velicky, J. Masek, V. Chab, B.A. Orlowski, Acta Phys. Pol. A69 (1986) 1059
/32/   J. Dreyhsig, U. Stutenbäumer, H.-E. Gumlich, J. Cryst. Growth 101 (1990) 443
/33/   R. Weidemann, B. Burmester, H.-E. Gumlich, Ch. Jung, T. Kleemann, T. Kreitler,
       A. Krost, H.-U. Middelmann, U. Becker, M. Kupsch, S. Bernstorff
       J. Cryst. Growth 101 (1990) 916
/34/   B.E. Larson, K.C. Hass, H. Ehrenreich, A.E. Carlsson, Phys. Rev. B 37 (1988) 4137
/35/   S.H. Wei, A. Zunger
       Phys. Rev. Lett. 56 (1986) 2391
       Phys. Rev. B 35 (1987) 2340
/36/   F.M.F. de Groot, J.C. Fuggle, B.T. Thole, G.A. Sawatzky, Phys. Rev. B 42 (1990) 5459
/37/   C. Kunz in Photoemission in Solids II, L. Ley, M. Cardona (Eds.), Springer Verlag 1979
/38/   T. Kleemann, Diplomarbeit, Technische Universität Berlin, 1990
/39/   B.T. Thole, R.D. Cowan, G.A. Sawatzky, J. Fink, J.C. Fuggle,
       Phys. Rev. B 31 (1985) 6856

ELECTRONIC STRUCTURE OF MONODISPERSED DEPOSITED PLATINUM CLUSTERS

P. Fayet[a], W. Eberhardt[b], D. M. Cox, Z. Fu, R. Sherwood, D. Sondericker and A. Kaldor.

Exxon Research and Engineering Co., Route 22E, Annandale, NJ 08801, USA.
[a] Université de Lausanne, Institut de Physique Expérimentale, BSP-Dorigny, CH-1015 Lausanne, Switzerland.
[b] Institut für Festkörperforschung der Kernforschungsanlage, D-5170 Jülich, Germany.

SUMMARY:

The electronic core level and valence band structures of Pt_1 to Pt_6 clusters have been studied by photoemission excited by synchrotron radiation. The samples were produced from a mass selected beam of cluster ions deposited onto the SiO_2 layer of silicon wafers. Each sample contained clusters having the same number of atoms. From these spectroscopic results, we can conclude that such small clusters were not yet metallic.

INTRODUCTION:

The formation of bands in condensed matter has been an intriguing subject since Wilson [1] described them in terms of a collective behaviour of a large number of electrons interacting with a periodic lattice. Band theory allows us to understand the sharp distinction between metals and non-metals: when the material is a metal, one or more bands are partly occupied, while in the case of a non-metal all bands are full or empty. The next step was done by Mott [2] who looked at the dynamical changes of bands when hydrogen-like atoms are put together to form a crystalline array with a lattice constant a. With an increase in a, at a given temperature, the material undergoes a metal-insulator transition which corresponds to a transition between a situation where bands overlap to a situation where they do not.

Instead of building a solid from N_a (Avogadro's number) atoms far apart, condensed to a solid lattice, we can look at the genesis of the formation of a solid atom per atom. By adding one building block after another, one studies how atomic constituents hold together to form clusters, one follows how the properties of aggregates change during the growth process. The growth is labeled by the integer number of units enclosed in the aggregates and this number is called the cluster size. When the size increases the aggregates show a richness of phase transitions occuring with well defined steps.

Most of the interesting properties of clusters arise from the transitions from bonds to bands as the cluster size increases. As the size increases the electrons become more and more delocalized over the cluster. Actually, the delocalization of valence electrons over a small drop of sodium occurs at a size as low as 10 atoms [3,4]. However, delocalization of electrons over all the atoms is not sufficient to make a cluster metallic. In the bulk, the properties are mainly determined by the density of states, while in the case of clusters, the discrete nature of electronic states plays a crucial role. The clusters have fewer conduction electrons than the bulk and the number of filled states up to the Fermi energy is small. The spacing between two adjacent conduction electron states is much larger. For a spherical particle of about 100 atoms, the level spacing is the inverse of the occupied levels. Comparing the energy spacing with the thermal energy $k_B T$, the electronic model for clusters is the following: although the conduction electrons may be delocalized over the entire cluster, a small cluster is insulating as long as the level spacing near the Fermi energy exceeds $k_B T$, while it is metallic when the spacing is smaller than $k_B T$. By changing the temperature, the clusters undergo a metal-insulator transition determined not only by the density of states but also by the statistics of the electron level distribution. This was first pointed out by Fröhlich in 1937 [5]. Later Kubo [6] and Kawabata [7] recognized the consequences of the discreteness of the spectrum on the properties of small particles.

A possible approach to study the electronic structure of clusters is to use photoelectron spectroscopy. The previous experimental investigations were carried out on small supported clusters of metallic elements grown by vapor nucleation on substrates [8-12], or on samples prepared from decomposition of organometallic precursors [13]. In such studies the clusters are always present in a size distribution with an average size which must be evaluated from different observations. In contrast this article presents valence- and core- level photoemission data of mass selected, monodispersed Pt_n clusters (n=1 to 6). The clusters were deposited from a beam of positively charged ion clusters onto the natural oxide layer of Si wafers.

EXPERIMENTAL DETAILS:

High currents of cluster ions were conveniently produced in a sputtering process by bombarding a metallic target with a fast ion beam. Combined with quadrupole mass spectrometry, the sputtering arrangement delivered continuous beams of size-selected cluster ions [15] Briefly, a Pt target was bombarded with Xe^+ ions. The ejected Pt_n^+ ions were mass selected with a quadrupole mass spectrometer and deposited onto a silicon wafer. The beam intensities were monitored during the deposition by using the Si wafer as a Faraday plate connected to the ground via an electrometer. Typical intensities were in the range of few nanoamperes for the larger measured clusters (Pt_6). The cluster beam energy was defined by the target bias voltage. At 10 V, the energy distribution, measured by the retarding potential technique, was found to be 10±5 eV. In order to limit the nucleation of the deposited clusters on the Si wafer surfaces, the total atomic coverage was kept at or less than 10^{14} cm^{-2}. All the depositions and the measurements were done at room temperature [16].

After the deposition was completed the samples were transported in a small UHV vacuum system (10^{-8} mbar) to the National Synchrotron Light Source (NSLS) at Brookhaven National Laboratory USA, where they were transfered into the UHV photoemission setup at 10^{-10} mbar of the U1 beam line [17]. The synchrotron radiation was monochromatized using the ERG monochromator installed on this line. The electrons were analysed by a commercial hemispherical electron energy analyser with 100 mm mean radius (Microscience HA 100). The overall resolution for the valence band spectra was better than 250 meV for a photon energy of 40 eV. The combined resolution of the monochromator and electron spectrometer was chosen to be 0.6 eV for the Pt 4f core levels taken at a photon energy of 280 eV. Simultaneously to the measurements, the Si $2p$ core levels of the substrate were taken and were used as an internal standard for the photon energy calibration. We could clearly observe both the clean Si $2p$ and the chemical shifted Si $2p$ emission from the oxide layer. From considerations of the mean free path of electrons at this kinetic energy, the oxide layer of the Si wafer is estimated to be about 15 Å thick.

RESULTS AND DISCUSSION:

The energies of the $4f_{7/2}$ levels for the platinum clusters from Pt_1 to Pt_6 are plotted in Fig.1 (a). The binding energy, measured relative to the Fermi level, clearly changes with cluster size. Compared with the bulk Pt metal value of 71.2 eV, the energy shift is quite large. The corresponding shift of the $4f_{7/2}$ level in the deposited atoms is about 2 eV and, in going from cluster size of 1 to 6 atoms, the shift shows a more or less smooth variation toward lower binding energies as can be expected from an intuitive molecular model of cluster growth.

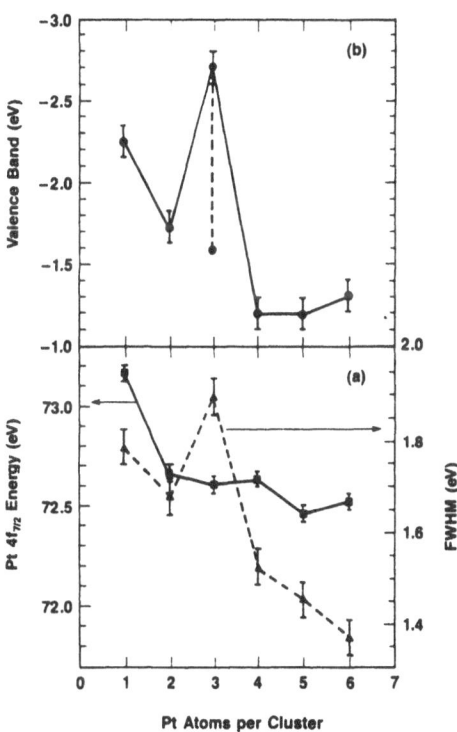

Fig.1. (a) The energy position relative to the Fermi energy and width (FWHM) of the Pt $4f_{7/2}$ core level lines, measured at $h\upsilon=280$ eV, as a function of the deposited cluster size. The binding energy of the Pt bulk metal is 71.2 eV and the width is less than 0.5 eV.
(b) The top of the valence band as a function of the cluster size measured at $h\upsilon=40$ eV. The values are plotted relative to the substrate Fermi level (5.1 eV [20]). The large width and the two onsets of the valence band at the size of 3 atoms are due to two different isomers for Pt_3. The figure is from [16].

The previous reported values of the $4f$ binding energy were taken from Pt clusters grown by the technique of nucleation induced by evaporation of the metal vapor onto the substrate [8,11,12,14]. Interestingly, the energy values presented here are significantly larger than the lowest coverage deposits obtained by this method. The closest value reported so far is for Pt particles embedded in amorphous SiO_2 [14], for which the core levels were measured to shift by 1.75 eV to higher binding energy at the lowest metal concentrations. However, none of the earlier studies had direct evidence that the particles studied were indeed small clusters of 6 atoms or less or single atoms as in our case. This supports the assumption that the majority of atoms and clusters survive the deposition from the gas phase as entities and that remain well dispersed onto the SiO_2 substrate [16].

From a surface science point of view, the deposition at 10^{-6} mbar (largely dominated by rare gas Xe from the sputter source) and the transportation of samples at 10^{-8} mbar for about 4 hours prior to insertion into the UHV spectrometer at 10^{-10} mbar, are not the most ideal conditions. Thus the binding energy shift presented above, or part of it, could be due to chemisorption of background gas as, for example, due to oxidation of the Pt clusters. This point was discussed earlier [16]. Briefly, the contamination of the samples can be ruled out except for hydrogen for the following reasons. Firstly, the chemical shifts for various Pt oxides and hydroxides are typically about 3 eV for the stoichiometric compounds [18]. This is much larger than the shift of 2 eV of the deposited atoms. Secondly, XPS studies of Pt catalysts show an oxide core level shift of about 2 eV [19], however, the particles consisted of more than 100 Pt atoms. Our largest cluster contains only 6 atoms and its core level shift is of 1.3 eV, indicating that it is not oxidized. Moreover, in the valence band measurements, certain features, which would indicate the presence of molecular adsorbates like CO, are absent [16].

Such energy shifts are not out of the range to be explained by purely a change in the final-state relaxation, but based on photoemission results alone, we cannot unambiguously differentiate between initial- and final-state effects. A change in the work function of the clusters can also be ruled out, since it is not identical with the shift measured for the top of the valence band as function of the cluster size.

Fig. 1 (a) shows also that the linewidth decreases with increasing cluster size. An exception is noted for the trimer, its fairly large width is attributed to the presence of two different isomers with slightly different binding energy. The observation that small particles exhibits a larger linewidth of the core level than the bulk Pt metal is in agreement with earlier results [8,9,10,12]. As explained in [16] the widening can be attributed to phonon broadening in addition to broadening induced by inhomogeneities in the SiO_2 sample.

On the SiO_2 substrates the cluster valence electron states show intensities in an energy range where substrate emission is nearly absent, since the top of the Si-oxide valence band emission is at 4.9 eV below the Fermi level [20]. The top region of the density of states of the clusters does not exhibit the characteristic step function of the metallic Fermi level, nor is it close to the region of the Fermi energy position. In spite of the possibility of hole localization which might account for a shift in energy of the valence band onsets [9,10], we conclude that the clusters in the size range investigated here are not yet metallic. The lack of any Fermi step yields a fairly reliable evaluation of the location of the top of the valence band states of the deposited clusters by linear extrapolation as is usually done for semiconductor materials. These values, obtained at 40 eV photon energy, are plotted in Fig. 1 (b). They show a strong size-dependent ionization potential (IP) as generally observed in studies of gas phase metal clusters [21]. The Pt_3 valence electron emission shows two onsets at 1.6 eV and at 2.7 eV below the Fermi energy. These features are due to the presence of two Pt_3 isomers on the SiO_2 surface: possibly a triangular one and a linear one. This assignment is also supported by the large width observed for the $4f$ core level of this cluster.

CONCLUSION:

The photoemission studies of Pt_1 to Pt_6 deposited on SiO_2 at room temperature show individual and discrete electronic structure which is characterized by the valence band energy onset and by the Pt $4f$ core level binding energy together with the corresponding linewidth. These findings confirm that each sample contained metal cluster systems with individual size. This establishes indirectly that clusters may survive the deposition and that they keep their identity without any subsequent aggregation on the SiO_2 surface. The valence band spectra do not present the step like Fermi function known for bulk metal. Consequently the clusters are not yet metallic. However, they may exhibit unique structural and chemical properties which should be further investgated by electron spectroscopy and by tunneling microscopy.

REFERENCES:

[1]   A. H. Wilson, Proc. R. Soc. Lond. A133, 458 (1931); A. H. Wilson, Proc. R. Soc. Lond. A134, 277 (1931).

[2]   N. F. Mott, Proc. Phys. Soc. A62, 416 (1949).

[3]   W. de Heer, K. Selby, V. Kresin, J. Masui, M. Wollmer, A. Chatelain and W. Knight, Phys. Rev. Lett. 59, 1805 (1987).

[4]   C. Bréchignac, P. Cahuzac, F. Carlier and J. Leygnier, Chem. Phys. 164, 433 (1989).

[5]   H. Fröhlich, Physica (Utrecht) 4, 406 (1937).

[6]   R. Kubo, J. Phys. Soc. Jpn. 17, 975 (1962).

[7]   A. Kawabata and R. Kubo, J. Phys. Soc. Jpn. 21, 1765 (1966).

[8]   M. G. Mason, Phys. Rev. B27, 748 (1983), and references therein.

[9]   S. B. DiCenzo and G. K. Wertheim, Comments Solid State Phys. 11, 203 (1985), and references therein.

[10] G. K. Wertheim and S. B. DiCenzo, Phys. Rev. B37, 844 (1988).

[11] A. Masson, B. Bellamy, Y. Hadj Romdhane, M. Che, H. Roulet and G.Dufour, Surf.   Sci. 173, 479 (1986).

[12] T. T. P. Cheung, Surf. Sci. 127, L129 (1983); 140, 151 (1984).

[13] E. W. Plummer, W. R. Salaneck and J. S. Miller, Phys. Rev. B18, 1673 (1978).

[14] V. Murgai, S, Raaen, M. Strongin and R. F. Garrett, Phys. Rev. B33, 4345 (1986).

[15] P. Fayet and L. Wöste, Spectrosc. Int. J. 3, 91 (1984); Surf. Sci. 156, 134 (1984); Z. Phys. D3, 177 (1986).

[16] W. Eberhardt, P. Fayet, D. M. Cox, Z. Fu, A. Kaldor, R. Sherwood and D. Sondericker, Phys. Rev. Lett. 64, 780 (1990); Physica Scripta, 41, 892 (1990).

[17] M. Sansone, R. Hewitt, W. Eberhardt and D. Sondericker, Nucl. Instrum. Methods Phys. Res., Sect. A266, 422 (1988).

[18] K. Duckers, K. C. Prince, H. P. Bonzel, V. Chab and K. Horn, Phys. Rev. B36, 6292 (1987), and references therein.

[19] W. Hoffman, M. Graetzel and J. Kiwi, J. Mol. Catal. 43, 183 (1987).

[20] F. J. Himpsel, F. R. McFely, A, Taleb IbraHimi, J. A. Yarmoff and G. Hollinger, Phys. Rev. B38, 6084 (1988).

[21] R. L. Whetten, D. M. Cox. D. J. Trevor and A. Kaldor, Phys. Rev. Lett. 54, 1494 (1985).

On the Application of Synchrotron Radiation for Electronic Structure Studies of Liquid Metals and Alloys

G. Indlekofer

LURE, Université Paris-Sud, 91405 Orsay, France

P. Oelhafen

Institut für Physik, Universität Basel, Klingelbergstr. 82, CH-4056 Basel, Switzerland

Selected experiments are briefly discussed in order to demonstrate the advantages and problems of investigating liquid metallic surfaces by means of synchrotron radiation.

Introduction

The **interest of synchrotron radiation (SR) studies** of liquid metals is based on a variety of reasons. Among those are attempts for a better understanding of (i) the electronic structure of liquid matter, (ii) basic properties of liquid metals in comparison with their corresponding solid amorphous and crystalline phases, (iii) surface properties of liquids and (iv) atomic cohesion in the disordered state. There exist however a number of **problems associated with liquid metallic surfaces** which dramatically limit the number of experimental investigations of the electronic structure, such as the evaporation, the high temperatures needed, the diffusion of impurities to the surface, the chemical reactivity of metallic melts, as well as their chemical and mechanical stability. In liquid alloys there can be in addition severe surface segregation. Up to 1987, the few published photoelectron spectroscopy studies of liquid metallic systems have all been performed by means of conventional laboratory light sources (i.e. gas discharge lamps and Al K_α or Mg K_α radiation). These and later investigations focussed essentially on liquid noble metals [1–3], polyvalent metals [4–16], alkali metals [5,17] and selected alloys with tolerable evaporation rates like Au-Sn [18,19], Ga-Sn [20], Pd-Si [21] or Tl-Bi [22]. It has been demonstrated through the first systematic study of liquid polyvalent metals combining both XPS ($h\nu$ = 1253.6 eV) with UPS over an extended energy range (11.8 eV $\leq h\nu \leq$ 48.4 eV) [23] that the former studies suffered partly from undetected surface contamination, and essentially all either from the presence of UV satellites and/or from the limitation to low excitation energies ($hv \leq 21.2$ eV). Such considerations underline the primary **advantages in the use of SR**, i.e. its tunability and spectral purity. Other important characteristics of the SR are its polarization, the well defined pulsed time structure and the fact that is is a `clean' source, i.e. it is not accompanied by vacuum deterioration in the analysis chamber. First publications on the application of SR in the UV regime to metallic liquids appeared in 1988 [24] followed by few others [25–27]. A report on the remarkable progress in the theoretical - computer simulated description of liquid elements obtained recently can be found in [28] and references therein.

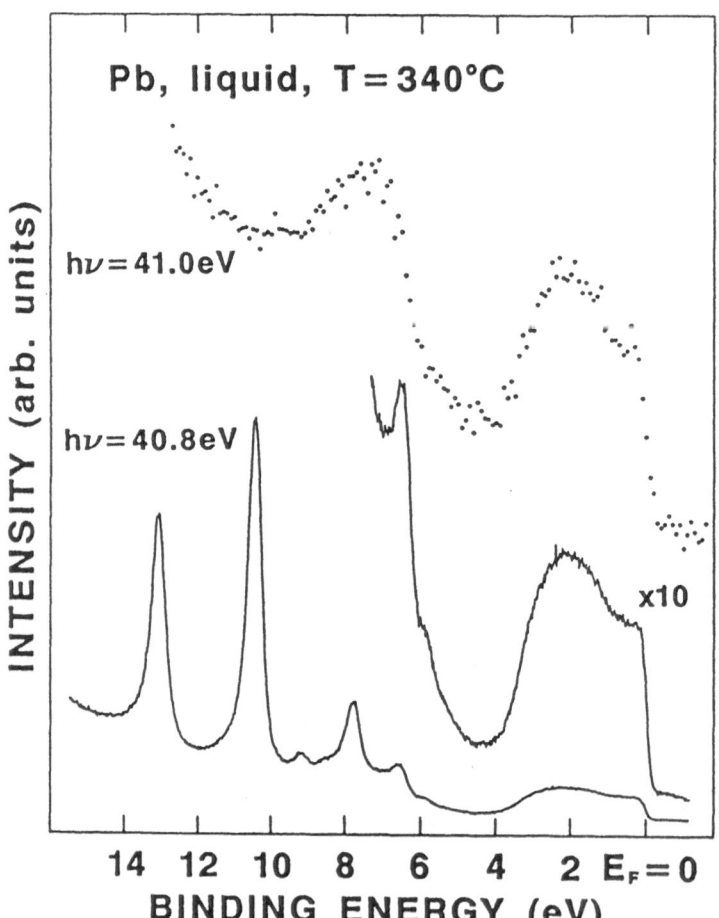

Fig. 1: Valence band photoelectron spectra of liquid lead excited by synchrotron radiation (hv = 41.0 eV) and by He II radiation (hv = 40.8 eV). For experimental details see [20, 26].

Discussion of selected experiments

For polyvalent metals the use of conventional UV sources did not allow the observation of the valence bands over their entire energy range. Reasons for this are the presence of the UV satellites, the large band widths (8 - 14 eV), the occurence of $X_{4,5}VV$ Auger transitions ($X = M, N, O$) due to weakly bound d states, and the ratio of the p to s photoionization cross sections [29] which prevents the study of the s contribution to the density of states by low UV energies ($hv < 25$ eV). This problem is best solved by the use of SR as demonstrated by the comparison given in Fig. 1 as an example. The spectrum obtained by SR resembles most directly the valence band of liquid lead which is split into two parts by a distinct energy gap. The two sub-bands consisting of states with essentially $6s$ and $6p$ character show up approximately at binding energies above 6 eV and below 3.8 eV respectively. The rise of the spectrum towards high binding energies originates from the smooth background and the influence of the analyzer transmission function. In contrast, the He II spectrum ($hv = 40.8$ eV) is heavily perturbed by satellite excitation ($hv_{sat} = 48.4, 51.0, 52.3$ and 53.0 eV). Only below 5 eV binding energy is the spectrum nearly identical to the one above. At high energies it is dominated by satellite excited $5d$ peaks (binding energies = 19.93 and 20.59 eV [30]) which screen the emission from the $6s$ band. It was shown [26] that even with low count rates, reasonable spectra can be obtained due to the excellent stability of the surfaces against recontamination once they have been cleaned properly [20].

In the photoemission valence band study of liquid AgCuGe [27] the excitation energy has been varied from 24 to 110 eV, approaching the Cooper minimum of the Ag $4d$ band photoionization cross section. As a consequence a direct discrimination between the d bands of Cu and Ag could be made, which otherwise greatly overlap in conventional photoelectron spectra. This allows the direct investigation of the local (i.e. atomically decomposed) density of states.

In a similar way the variation of the photon energy has been used in order to determine partial (i.e. angular momentum decomposed) densities of states, e.g. for polyvalent metals [15, 16]. In this case advantage has been taken from the huge variation of the relative s and p band photoionization cross sections even within the UV range of discharge lamps. The photoelectric yields for s and p states of a pure polyvalent metal turned out to be almost identical when the UV energy is around 40 eV or higher, whereas for $hv < 25$ eV the p band intensity exceeds that of the s band by at least an order of magnitude, in good qualitative agreement with theoretical predictions for atomic states [29].

Perspectives

The few examples mentioned above demonstrate already that major advantages are achieved from the application of SR. More progress can be expected by applying constant final state spectroscopy and constant initial state spectroscopy with their ability to investigate the unoccupied density of electronic states [31]. These techniques require a tunable light source and are therefore accesible only by SR. In addition the serious problem of evaporation, which is in general harmful to the spectrometer, can be overcome at least quantatively by the use of more brilliant sources (e.g. synchrotrons with undulators), by properly making use of the beam time structure of the synchrotron beam, e.g. by triggered heating by Laser irradiation, and by the design of spectrometers which allow a reasonable removal of the deposited metallic layers. The use of Laser heating along with sample cooling can in addition solve the problem of reactivity since in this case the liquid metal can be kept within a crucible consisting of its own solid phase. It can be concluded that major progress in the knowledge of the properties of liquid matter can be expected from the more frequent application of synchrotron radiation.

Acknowledgement:
This work has been supported by the Swiss National Science Foundation.

References:
[1]   D.E. Eastman; Phys. Rev. Lett. **26**, 1108 (1971)

[2]   G.P. Williams and C. Norris; J. Phys. F **4**, L175 (1974)

[3]   G.P. Williams and C. Norris; Phil. Mag. **34** 851 (1976)

[4]   R.Y. Koyama and W.E. Spicer; Phys. Rev. B **4**, 4318 (1971)

[5]   J.E. Enderby; in *Liquid Metals,* ed. by S.Z. Beer (Marcel Dekker, New York 1972) p. 585

[6]   P. Cotti, H.-J. Güntherodt, P. Munz, P. Oelhafen and J. Wullschleger; Solid State Commun. **12**, 635 (1973)

[7]   C. Norris, D.C. Rodway and G.P. Williams; in *The Properties of Liquid Metals,* ed. by S. Takeuchi (Taylor & Francis, London 1973) p. 181

[8]   C. Norris, D.C. Rodway and G.P. Williams; J. Physique C4, 61 (1974)

[9]   C. Norris and J.T.M. Wotherspoon, J. Phys. F **6**, L263 (1976)

[10]   P. Oelhafen; PhD thesis, ETH Zürich (1976)

[11]   Y. Baer and H.P. Myers; Solid State Commun. **21**, 833 (1977)

[12]   P. Oelhafen, U. Gubler and F. Greuter; in *Electrons in Disordered Metals and at Metallic Surfaces,* ed. by P. Phariseau, B.L. Gyorffy and L. Scheire (Plenum Publ. Corp., New York 1979) p.337

[13]   J.T.M. Wotherspoon, D.C. Rodway and C. Norris, Phil. Mag. B **40**, 561 (1979)

[14]   M.K. Gardiner, D. Colbourne and C. Norris, Phil. Mag. B **54**, 133 (1986)

[15]   G. Indlekofer, P. Oelhafen, R. Lapka and H.-J. Güntherodt; Z. Phys. Neue Folge **157**, 465 (1988) (Oldenbourg, München 1987)

[16]   P. Oelhafen, G. Indlekofer and H.-J. Güntherodt; Z. Phys. Neue Folge **157**, 483 (1988) (Oldenbourg, München 1987)

[17]   G. Indlekofer and P. Oelhafen; J. Non-Cryst. Solids **117/118**, 340 (1990)

[18]   T. Ichikawa; Phys. Stat. Sol. (a) **32**, 369 (1975)

[19]   G. Indlekofer, A. Pflugi, P. Oelhafen, H.-J. Güntherodt, P. Häussler, H.-G. Boyen and F. Baumann; J. Mater. Sci. Engineering **99**, 257 (1988)

[20]   G. Indlekofer, P. Oelhafen and H.-G. Güntherodt; in *MRS Symposia Proceedings* **83**, ed. by D.M. Zehner and D.W. Goodman (1987) p. 75

[21]   A. Pflugi, G. Indlekofer and P. Oelhafen ; J. Non-Cryst. Solids **117/118**, 336 (1990)

[22]   P. Häussler, G. Indlekofer, H.-G. Boyen, P. Oelhafen and H.-G. Güntherodt; J. Mater. Sci. Engineering, in press

[23]   G. Indlekofer and P. Oelhafen; in *Disordered Systems and New Materials,* ed. by M. Borrisov, N. Kirov and A. Vavrek (World Scientific Publishing Co., Singapore, 1989) p. 707

[24]   A. Kakizaki, M. Niwano, H. Yamakawa, K. Soda, S. Suzuki, H. Sugawara, H. Kato, T. Miyahara and T. Ishii; J. Phys. F: Met. Phys. **18**, 2617 (1988)

[25]   A. Kakizaki, M. Niwano, H. Yamakawa, K. Soda, T. Ishii and S. Suzuki; J. Non-Cryst. Solids **117/118**, 417 (1990)

[26]   G. Indlekofer, A. Pflugi, P. Oelhafen, D. Chauveau, C. Guillot and J. Lecante; J. Non-Cryst. Solids **117/118**, 351 (1990)

[27]   P. Oelhafen, A. Pflugi and G. Indlekofer; J. Non-Cryst. Solids **117/118**, 267 (1990)

[28]   J. Hafner and W. Jank, Phys. Rev. B 42, 11530 (1990)

[29]   J.J. Yeh and I. Lindau; At. Data and Nucl. Data Tables **32**, 1 (1985)

[30]   G. Indlekofer, PhD thesis, Universität Basel (1987)

[31]   see: *Handbook on Synchrotron Radiation Vol. 2,* ed. by G.V. Marr (North Holland, 1987) p. 599 and references given therein

Synchrotron Radiation: Selected Experiments in Condensed Matter Physics, Monte Verità, © Birkhäuser Verlag Basel

THEORY OF X-RAY ABSORPTION

The Crystal-Field Multiplet approach

F.M.F. de Groot

Research Institute for Materials, University of Nijmegen

Toernooiveld, NL-6525 ED Nijmegen, The Netherlands

1   Introduction

This contribution concerns the interpretation of x-ray absorption spectra for the determination of the electronic structure. This subject has gained much impetus in recent years as a direct consequence of the experimental progress in the field of soft x-ray monochromators, like the SX700 plane-grating-monochromators [1] and the DRAGON monochromator, based on a cylindrical-grating [2]; both monochromators reaching up to 1:10.000 resolution in the soft x-ray range. This stimulated the development of many new fields one of which is 2p x-ray absorption ($L_{2,3}$-edges) of the 3d transition-metal compounds. The high resolution, combined with the relative long lifetime of the excited states, result in extremely sharp spectra with complex shapes. In the process of explaining these shapes it was immediately clear that the role of the core-hole was of central importance, like in the case of 3d x-ray absorption spectra (or $M_{4,5}$-edges) of the Rare-Earths. The experiments on the $M_{4,5}$-edges in the sixties [3] were, for some early Rare-Earths, partially explained with an atomic multiplet approach, developed in the early seventies [4]. This approach was improved and generalized to all Rare-Earths by Thole et al. and good agreement was found with experiment [5]. Although the atomic multiplet results show some likeness with the $L_{2,3}$-edges of the 3d transition-metal compounds, solid state effects modify the spectra to such extend that detailed analysis calls for a change in the approach. In the following we will in short give an overview of the some models to simulate the x-ray absorption cross section.

2   The x-ray absorption cross section

The study of x-ray absorption for electronic structure determination is concerned with two basic interactions: The electron-electron interactions in the material of investigation and the electron-photon interaction, i.e. the interaction of x-rays with matter. In the, generally valid, single photon approximation the x-ray absorption cross section is given as a transition matrix element (squared) times a delta function:

$$\sigma(E_{h\nu}) \sim \sum_{f} |\langle \phi_i | X | \phi_f \rangle|^2 \, \delta(E_i + E_{h\nu} - E_f) \qquad (1)$$

$E_{h\nu}$, E_i and E_f are respectively the energy of the photon, the initial state and the final state. X is the perturbation acting on the system, for which we will use the dipole approximation [6].

2.1 Electronic structure calculations

To calculate the x-ray absorption cross section it is thus neccesary to find the initial and final state energies and wave-functions. The usual approach to this problem in the solid state is to transform the general many-electron wave-function to a single particle wave-function by using the Density Functional Theory [7]. Whether one uses a bandstructure approach or a multiple scattering approach should make no difference, given that all further approximations, dependent on the specific method of calculation, do not introduce large errors [8]. In the Density Functional approach to x-ray absorption, the initial state wave-function is that of a core state (e.g. $2p$), from which transition are posible to empty states, which due to the dipole selection rule have either s or d character. This transforms equation (1) to the energy-dependent matrix element $M(E)$ times the projected density of states ($\mathcal{P}_{s,d}$):

$$\sigma(E) \sim\mid M_d(E) \mid^2 \cdot \mathcal{P}_d^*(E) + \mid M_s(E) \mid^2 \cdot \mathcal{P}_s^*(E) \tag{2}$$

In this equation $\mathcal{P}^*$ is used to stress the presence of the core hole. Implicit in this approach is the neglect of all multipole interactions of the core hole with the valence states; the core hole is included only as an extra attractive potential. However because the task of accurately and self-consistently calculating the final state density-of-states is, often, still a too complicated problem, further approximations are often allowed for. These includes the complete neglect of the core hole, in which case the XAS spectral function simply relates to the ground state densiy-of-states. Also the energy dependence of the matrix elements is often neglected. Futhermore it is difficult to include ground state multiplet effects ('orbital polarization') in the DFT calculations [9].

The approximations made in this approach therefore are, except from calculational approximations which differ between the various methods:

- The errors introduced by the Density Functional, single particle, approach to describe the electronic states.

- The neglect of multipole core hole effects.

- (in some cases) The complete neglect of core hole.

- (in some cases) The neglect of energy-dependence of the matrix elements.

- (in most cases) The neglect of 'orbital polarization'

The Density Functional approach is often rather successfully applied to x-ray absorption (near edge) spectra, including the K edges of the 3d transition metals, the K edge of oxygen. For historical reasons, particularly the success of scattering theory for the explanation of the extended x-ray absorption fine structure (EXAFS), the hard x-ray absorption edges are described normally with multiple scattering theory. In the soft x-ray region, band structure calculations are prefered because the computational problems, especially those related to self consistent solutions, are in general treated more accurately. However the multiple scattering method is again easier to adapt to e.g. surface problems. As stated before, in principle both methods if applied without further approximations should result in the same XAS spectra [8].

2.2   Atomic Multiplet Approach

We now turn to a case where the Density Functional approach fails completely: The $M_{4,5}$ edges of Rare Earth materials. The (multipole) interaction of the created 3d core hole with the partly filled 4f level is so strong that it completely dictates the shape of the spectrum. The most obvious approach therefore is to neglect all solid state effects and to calculate the $M_{4,5}$ x-ray absorption spectra as taking place in an atom. A simple means to calculate the actual spectral shape is to use the Hartree-Fock approximation to calculate both initial and final state atomic wave functions and coupling both again with the appropiate matrix elements. The x-ray absorption cross section can thus be written as:

$$\sigma(E_{h\nu}) \sim \sum_f | \langle 4f_{HR}^N \mid X \mid 3d^9 4f^{N+1} \rangle |^2 \, \delta[E(4f_{HR}^N) + E_{h\nu} - E(3d^9 4f^{N+1})] \qquad (3)$$

It was found that to compare the Hartree-Fock results to atomic experimental data, one had to reduce the Slater Integrals to approximately 80% of their calculated value. This, crude, approximation was followed for the successfull calculation of the $M_{4,5}$ edges of the Rare Earths [5]. No configuration interaction effects are taken into account explicitly.

The approximation made in this Atomic Multiplet approach can be summarized as:

- The complete neglect of all solid state effects.

- Crude calculation of energy levels: Hartree-Fock with emperical reduction to 80% of Slater Integrals.

2.3   Crystal Field Multiplet approach

A situation intermediate between these two extreme approaches is presented by the $L_{2,3}$ edges of the 3d transition metal oxides. As for the Rare Earths we are confronted with a partly filled, correlated, shell and a polarised core hole. On the other hand in the solid state the 3d-band is far less atomic in character, and in fact the dominating object of study for the 3d transition metal oxides are the interactions between the partly filled 3d-band and the full oxygen 2p-band, with all its consequences for odd optical, electrical and magnetical behaviour. The $L_{2,3}$ edges are dominated by transitions to the empty part of the 3d-band. At higher energies (some eV above E_f) transitions to the '4s-band' occur, but the intensity is much lower.

The Density Functional approach should in principle reproduce the general features of the spectrum, but because the multipole interactions of the 2p core hole with the 3d valence band can not be neglected, the shape of the '3d-band' is completely modified. Therefore the best way to calculate the (near) edge structure is to include the multipole core hole interactions explicitly and to take the solid state effects into account as a perturbation only.

The starting point is again a Atomic Multiplet calculation of the $3d^N \rightarrow 2p^5 3d^{N+1}$ transition. The solid state is then included as the modification of the electric field from spherical to the appropiate situation in the crystal. The splitting of the 3d-band in transition metal compounds is dominated by the cubic crystal field: The 5-fold degenerate d-orbitals are split into a t_{2g} (d_{xy}, d_{xz}, d_{yz}) and an e_g (d_{z^2}, $d_{x^2-y^2}$) manifold. This cubic crystal field effect can be included in the Atomic Multiplet calculations by transforming the wave-functions to cubic symmetry. The strength of the cubic crystal field is fitted to experiment. Again hybridization effects are not considered explicitly.

The criterion which approach is best suited for a specific x-ray absorption spectrum comes down to a comparison of the dispersional broadening of a band with the magnitude of the atomic multipole couplings (the Slater Integrals) of the core hole with the valence electrons. For $M_{4,5}$ edges of Rare Earths and the $L_{2,3}$ edges of the 3d transition metals the multipole interactions are in the order of respectively 8 and 5 eV (For F_{pd}^2), much bigger than the dispersional broadenings. For the K edges of the 3d transition metals and the $L_{2,3}$ edges of the Rare Earths the core hole Slater Integrals are small (less than 1 eV) and can be neglected [10].

3   Results of the Cubic Crystal Field approach

The results of the Crystal Field Multiplet approach for 3d transition metal compounds are treated in detail in Ref. [11]. As an example I will in short discuss the 2p x-ray absorption spectrum of MnF_2. The theoretical spectrum is generated in the following way: Mn in MnF_2 is assumed to be $d^5 Mn^{2+}$. The 3d-3d multipole interactions (F_{dd}^2, F_{dd}^4) form the ^6S ground state (Hunds rule). From this ground state transitions are made to all possible $2p^5 3d^6$ states. The resulting line spectrum is shown in Fig. 1. The cubic crystal field does not split the ground state in case of d^5. Therefore the only effect of the cubic crystal field is a splitting of the states in the final state. This in contrast to e.g. $3d^3$, where the ^4F ground state is split in $^4A_2$, $^4T_1$ and $^4T_2$. Fig. 2. shows the effect of an increasing cubic crystal field for Mn^{2+}. Comparison to experiment results in the optimized fit for a cubic crystal field strength of 0.75 eV. Fig. 3. shows the MnF_2 2p x-ray absorption spectrum compared to this calculation [12].

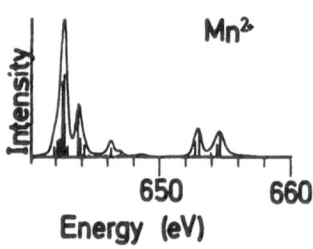

Figure 1: $d^5 Mn^{2+}$ atomic multiplet

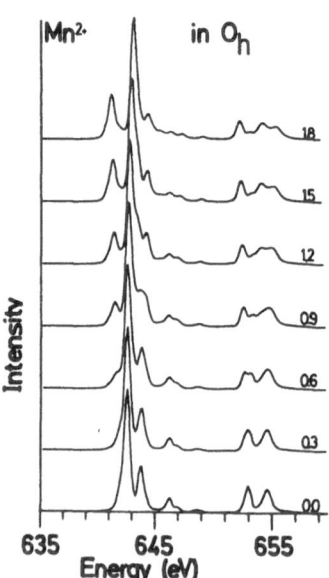

Figure 2: Effect of the cubic crystal field (given in eV).

Figure 3: XAS-spectrum (E) compared with crystal field calculation (T).

3.1   Further developments and limitations

The crystal field multiplet approach can be extended to lower symmetries. If the distortion from cubic cq octahedral symmetry is small the effect will be difficult to detect. However for cases where large distortions occur clear deviations from the cubic spectra are to be expected. This can be accounted for by projecting to lower symmetry, using a similar approach as the projection from spherical to cubic. This was studied in detail for the CaF_2 on $Si(111)$-system [13].

In contrast to the overall x-ray absorption spectrum, which is determined mostly by final state effects and which is not very sensitive to small splittings in the initial state, polarization dependent spectra are highly sensitive to small splittings in the initial state. These small splittings will be a combination of small distortions from cubic symmetry, 3d spin-orbit coupling and possible magnetic splittings. A clear case of polarization dependence originating from crystal field effects (and below T_c also from magnetic effects) has been demonstated for $BaCoF_4$ [14]. Polarization dependent effects ('dichroism') have been extensively used to study magnetic materials [15]. An implicit assumption in the crystal field model is that the electronic configuration of a transition metal is determined by the 3d-count only, i.e. that V^{2+} will relate to d^3, etc. However in general the ground state will be formed by a mixture of two or more configurations. A d^N state will mix with a $d^{N+1}\underline{L}$ state in which a hole is created in the ligand band. These hybridization effects can in principle be included, but much work has still to be done following this line 16]. The success of the crystal field multiplet model shows that in many cases the ground state is sufficiently described by its main configuration, as far as x-ray absorption is concerned.

[1] H. Petersen, *Nucl.Instrum.Methods A* **246**, 260, (1986).

[2] C.T. Chen *Nucl.Instrum.Methods A* **256**, 595, (1987).

[3] V.A. Fomichev, T.M. Zimkina, S.A. Gribovskii and I.I. Zhukova, *Soviet Physics, Solid State* 9, 1163, (1967); R. Haensel, P. Rabe and B. Sonntag, *Solid State Comm.* 8, 1845, (1970).

[4] J.L. Dehmer, A.F. Starace, U. Fano, J. Sugar and J.W. Cooper, *Phys.Rev.Lett.* **25**, 1521, (1971); A.F. Starace *Phys.Rev.B.* **5**, 1773, (1972); J. Sugar *ibid.*5, 1785, (1972); J.L. Dehmer and A.F. Starace *ibid.*5, 1792, (1972).

[5] B.T. Thole, G. van der Laan, J.C. Fuggle, G.A. Sawatzky, R.C. Karnatak and J.-M. Esteva, *Phys.Rev.B.* **32**, 5107, (1985).

[6] C.-O. Almbladh and L. Hedin, *Handbook on Synchrotron Radiation Vol. I (ed. E.E. Koch), Ch.8*, (1983).

[7] P.C. Hohenberg and W.Kohn, *Phys.Rev.B.*, **136**, 864, (1964); W. Kohn and L.J. Sham *Phys.Rev.A.* **140**, 1133, (1965); O. Gunnarson and R.O. Jones *Rev.Mod.Phys.* **61**, 689, (1989).

[8] R. Natoli and M. Benfatto, *J.Phys.Coll.C8*, **47**, 11, (1986).

[9] Olle Eriksson, Lars Nordström, Anna Pohl, Lukas Severin, A.M. Boring and Björe Johansson, *Phys.Rev.B.*, **41**, 11807, (1990).

[10] P.J.W. Weijs et al. *Phys.Rev.B.*, **41**, 11899, (1990).

[11] F.M.F. de Groot, J.C. Fuggle, B.T. Thole and G.A. Sawatzky, *Phys.Rev.B.* **41**, 928, (1990) and *Phys.Rev.B.* **42**, 5459, (1990).

[12] A series of 2p x-ray absorption spectra of manganese compounds with different crystal field strengths were measured. They were simulated with the crystal field multiplet calculations from which the crystal field strengths could be determined; S.P. Cramer, F.M.F. de Groot, Y. Ma, C.T. Chen, F. Sette, C.A. Kipke, D.M. Eichhorn, M.K. Chan, W.H. Armstrong, E. Libby, G. Christou, S. Brooker, V.McKee, O.C. Mullins and J.C. Fuggle, to appear in *J.Am.Chem.Soc.*

[13] F.J. Himpsel, U.O. Karlsson, A.B. McLean, L.J. Terminello, F.M.F. de Groot, M. Abbate, J.C. Fuggle, J.A. Yarmoff, B.T. Thole and G.A. Sawatzky, *Phys.Rev.B.* **43**, to appear 25-3-1991.

[14] B. Sinkovic et. al, to be published.

[15] J.B. Goedkoop, B.T. Thole, G. van der Laan, G.A. Sawatzky, F.M.F. de Groot and J.C. Fuggle, *Phys.Rev.B.* **37**, 2086, (1988); J.B. Goedkoop, *thesis University of Nijmegen*, (1989); see also the contributions of G. Schutz and M. Altarelli (this proceedings).

[16] see e.g. J. Zaanen, G.A. Sawatzky, J. Fink, W. Speier and J.C. Fuggle, *Phys.Rev.B.* **32**, 4905, (1985) and A. Kotani, H. Ogasawara, K. Okada, B.T. Thole and G.A. Sawatzky, *Phys.Rev.B.* **40**, 65, (1989). For charge conserving spectroscopies, like XAS, the core hole will not change the mixing a lot, thereby surpressing satellites. Because of this an alternative approach to account for hybridization is to simply reduce the Slater Integrals (see e.g. D.W. Lynch and R.D. Cowan, *Phys.Rev.B.* **36**, 9228, (1987).

Synchrotron Radiation: Selected Experiments in Condensed Matter Physics, Monte Verità, © Birkhäuser Verlag Basel

Partial densities of states of LaNi$_5$ measured with x-ray photoelectron diffraction

A.Stuck, D.Naumović, U.Neuhaus, J.Osterwalder, L. Schlapbach

Institut de Physique, Université de Fribourg, CH-1700 Fribourg (Switzerland)

We decompose the valence band photoelectron spectrum of LaNi$_5$ (11-20) into its La and Ni contributions. This is achieved by comparison of valence band photoelectron diffraction with La 5p and Ni 3p core level photoelectron diffraction. The well-known satellite at 6 eV binding energy in LaNi$_5$ valence band spectra is clearly attributed to Ni photoemission. In accordance with calculations we find the La VB emission centered around 1.4eV binding energy while the Ni 3d band has a maximum at 0.8 eV below the Fermi level.

As recently shown for AuCu$_3$(001), it is possible to exploit differences in the angular differential cross sections of each constituent at fixed energy in order to unfold the x-ray photoelectron (XPS) valence band (VB) spectra of alloys into the valence band partial density of states (PDOS).[1] These angular cross section variations can be measured very accurately by analyzing x-ray photoelectron diffraction (XPD) of shallow core levels in the same crystal. No comparison with properties of pure metals is needed, since the core level diffraction patterns of the alloy serve as fingerprints for the constituents. The decomposition is independent of theoretically determined parameters and yields the PDOS and the ratio of the VB to the core-level cross sections for each element. In the present paper this method is applied to LaNi$_5$. In contrast to the elements of AuCu$_3$ Ni and La have open d-shells and the VB photoemission spectra of LaNi$_5$ show a satellite around 6 eV binding energy, which is also found in the spectra of pure Ni.[2]

The valence bands and low binding-energy core levels of simple metals have very similar diffraction patterns, as has recently been shown by Osterwalder et al. for Al(001).[3] A localization of the photoemission hole state has been inferred to give rise to essentially the same scattering and diffraction of the photoelectrons originating from extended valence states as those from localized core levels. This similarity is expected to hold also for the VB of transition and noble metals [4,5], at least as long as direct transitions are not important. In LaNi$_5$, the high kinetic energy (large k-vector) favours indirect transitions and little angular variations due to direct transitions are expected.[5] Thus, the whole Brillouin zone is sampled. Neglecting the emission of valence s electrons,[6] it can be assumed, that the number of photoemitted VB electrons, I^{vb} (θ,ϕ,E), measured at a given binding energy E at angles θ and ϕ, is a linear combination of Ni 3d and La 5d contributions:

$$I^{vb}(\theta,\phi,E) = \sum_{i=La,Ni} I_o g_i^{vb}(\theta,\phi)\, \sigma_i^{vb}\, \frac{\rho_i^{vb}(E)}{n_i^{vb}} \tag{1}$$

Here, the angular photoelectron distribution from the i'th element (i=La,Ni) is described by a diffraction function $g_i^{vb}(\theta,\phi)$, containing the angular dependence, I_o is the photon intensity times the instrumental response, σ_i^{vb} the total cross section of the VB d states of La or Ni, $\rho_i^{vb}(E)$ is the photoelectron energy distribution curve (PEDC) of the i'th element. n_i^{vb} is the number of d electrons of the i'th element per unit cell. Likewise we can write for the energy-integrated intensities of the core level peaks (defined as $S_i^{core}(\theta,\phi) = \int I_i^{core}(\theta,\phi,E)dE$):

$$S_i^{core}(\theta,\phi) = I_o g_i^{core}(\theta,\phi)\, \sigma_i^{core} \tag{2}$$

In the case of Al(001),[3] the relative difference in the wavelengths between Mg Kα excited core and valence-band electrons is about 4.6%, while it lies between 0.7 and 2.8% for LaNi$_5$. In both cases, such small differences apparently have little effects on the diffraction patterns, as shown in Fig.1. We can thus assume that the diffraction functions of the core and VB electrons are equal and therefore neglect any energy dependence of $g_i^{vb}(\theta,\phi)$, as well as initial and final state angular momentum effects,[1] i.e: $g_i^{vb}(\theta,\phi) = g_i^{core}(\theta,\phi)$. Substituting Eq.2 into Eq.1 then yields:

$$I^{vb}(\theta,\phi,E) = \sum_{i=La,Ni} S_i^{core}(\theta,\phi)\, \frac{\sigma_i^{vb}}{\sigma_i^{core}}\, \frac{\rho_i^{vb}(E)}{n_i^{vb}} \tag{3}$$

and, after integrating over all energies $S^{vb}(\theta,\phi) = \int I^{vb}(\theta,\phi,E)dE$, to

$$S^{vb}(\theta,\phi) = \sum_{i=La,Ni} S_i^{core}(\theta,\phi) \frac{\sigma_i^{vb}}{\sigma_i^{core}} \qquad (4)$$

Measurements of the energy-integrated core level and VB intensities for at least N different directions (N=number of elements) thus yield a set of linear equations (Eq.4), which can be solved for the ratios of the cross sections. Given these N ratios and the energy resolved VB spectra, a second set of linear equations (Eq.3) at each energy channel remains which can be solved for the PEDC of the two elements. In the XPS regime both results, the cross section ratios and the PEDC, are independent of experimental parameters such as the photon intensity and the instrumental response which cancel. Furthermore, the study of dilute systems is only limited by the statistical error of the measurments. In LaNi$_5$ N=2 and the PEDC was calibrated by choosing $n_{La}^{vb}=1$ electron/cell for La and $n_{Ni}^{vb} = 44.85$ electrons/cell for Ni. In the XPS regime, the PEDC are often similar to the PDOS.[2] They do however include final state effects, like for example satellites. These effects are small when the d-shell is filled but can be important in systems with open d-shells as is the case in La and Ni.[2,11] Furthermore, any energy dependence of the VB cross sections does affect the PEDC as well as angular variations of the VB cross sections due to the symmetry of the d-shells.[7]

All experiments were performed with MgKα (hω=1253.6 eV) radiation and an energy resolution of about 1.2 eV. The experimental setup is described elsewhere.[3] The sample was oriented better than 0.5° and cleaned in situ by repeated cycles of 1kV Ar+ sputtering and subsequent flashing up to 950 K, until, at room temperature, sharp LEED spots showed a well ordered surface. This preparation minimized surface segregation of La. The stoichiometry was determined as Ni:La = 4.5 ± 0.5. The oxygen coverage was below 50% of a monolayer (ML), while the carbon contamination was constant at about 60% ML. All data presented were measured at a fixed polar angle θ=30° with respect to the surface normal. The data are thus rather bulk sensitive and the influence of surface contamination is minimized. All energy spectra were numerically corrected for MgKα satellites and a Shirley-type background was assumed. Linear background substraction procedures were also applied with no significant change in the form and intensity of the PEDC's. To achieve better statistics, the VB and core level spectra were measured at 3 different angles, as shown in Fig.1. Consequently the systems of linear equations Eq.3 resp. Eq.4 were overdetermined. The solutions, found by a least square method, are shown in Fig.2.

The Ni PEDC has a maximum at 0.8 eV binding energy. Its height and position correspond well with the calculations by Gupta.[8] In addition there is a satellite around 6 eV which is also found in XPS spectra of pure Ni.[9] As expected, this satellite can not be seen in the PEDC of La and thus is an intrinsic property of photoemission from Ni. The La PEDC is shifted towards higher binding energy with respect to Ni in $LaNi_5$ (Fig.2) and pure La.[10] It has a single peak around 1.4 eV which is asymmetric, falling to zero at 8 eV. In agreement wiht our measurments the calculations predict a d-band maximum around 1.6 eV.[8] However, a sharp peak around 2.5 eV, is predicted theoretically but does neither appear in our data nor in UPS measurments by Schlapbach.[11]

In summary, we have separated the La and Ni contributions of $LaNi_5$ into its La and Ni PEDC's. The satellite at 6 eV binding energy is clearly attributed to Ni photoemission and the maximum of the La PEDC is shifted towards higher binding energy with respect to Ni in accordance with calculations.

Acknowledgments:

We would like to mention the skillfull technical assistance of F. Bourqui, O.Raetzo and H. Tschopp. We are in debt to T. Greber who has made many holpfull suggestions. This work was supported by the Swiss National Science Foundation and the Swiss National Energy Foundation.

References:

[1]   A. Stuck, J. Osterwalder, T. Greber, S. Hüfner, L. Schlapbach, to be published
[2]   S.Hüfner,Photoemission in Solids II, Springer NY,Ed. L.Ley,
      M.Cardona,(1979),p.173 ff
[3]   J. Osterwalder, T. Greber, S. Hüfner, L. Schlapbach, Phys. Rev. Lett. 64, 2683, (1990)
[4]   R.J. Baird, C.S. Fadley, L.F. Wagener, Phys. Rev. B 15, 666 (1977)
[5]   R.C. White, C.S. Fadley, M.Sagurton, Z. Hussain, Phys. Rev. B 34,5226 (1986)
[6]   J.J.Yeh, I. Lindau, At.Data and Nucl. Tabl. 32, 1 (1985)
[7]   F.R. McFeely, J. Stöhr, G. Apai, P.S. Wehner, D.A. Shirley, Phys.Rev.B 14, 3273 (1976)
[8]   M. Gupta, J.Less-Common Met., 130, 219 (1987)
[9]   S. Hüfner, G.K. Wertheim, Physics Lett. 51A, 299 (1975)
[10] M.Campagna, Photoemission in Solids II, Ed.L.Ley,M.Cardona, Springer NY(1979), p. 217 ff
[11] L.Schlapbach, Solid State Comm. 38, 117 (1981)

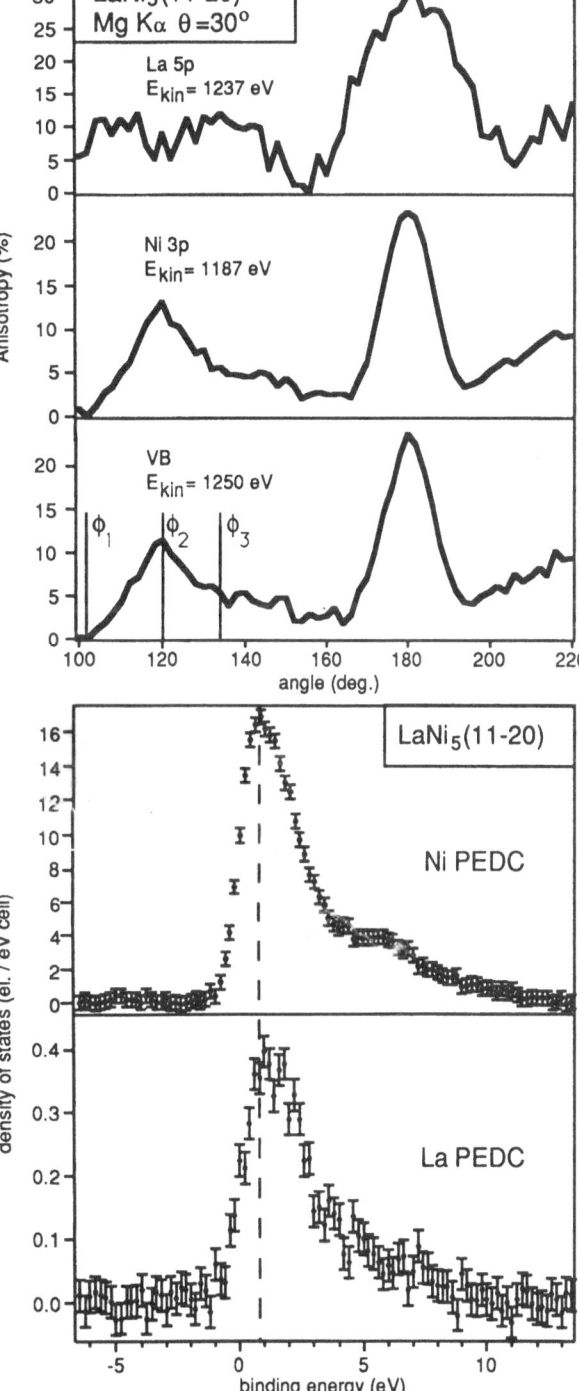

Fig.1
Comparison between La 5p, Ni 3p and VB diffraction patterns. The Ni 3p and VB patterns are very similar. The anisotropy is a measure of the strength of the diffraction. It is defined as (I-Imin)/Imax. Here I is the intensity of the diffraction at a given angle and Imin, Imax are the minimum resp. maximum intensities of a particular scan. The vertical lines indicate the 3 directions where energy scans for further analysis were taken.

Fig.2
VB partial photoelectron distribution curves for La and Ni. The Ni satellite around 6 eV is clearly seen and does not appear in the La PEDC. Ni has a considerable amount of electrons at the Fermi edge and a maximum at 0.8 eV, which is indicated by the vertical line.

Interfaces

Synchrotron Radiation: Selected Experiments in Condensed Matter Physics, Monte Verità, © Birkhäuser Verlag Basel

PROPERTIES OF SCHOTTKY BARRIER   FORMATION AS SEEN BY SYNCHROTRON RADIATION PHOTOEMISSION SPECTROSCOPY

R. Cimino *

Istituto di Struttura della Materia
Via E. Fermi 38
00044 Frascati, ITALY

1. SUMMARY

This short review is an attempt to present our current understanding of Schottky barrier formation as achieved using synchrotron radiation photoemission spectroscopy. No attempt is made to give an exhaustive report of all the studies available in the literature. Only a few representative experimental data will be analysed in detail when discussing actual capabilities of photoelectron spectroscopy to obtain insight into the problem of Schottky barrier formation.
This review is organized as follows: in section 2 the problem of Schottky barrier formation will be introduced; in sec. 3 a short overview will be presented of the most widely discussed theoretical models proposed to identify the driving forces governing the mechanism of metal-semiconductor junction formation; in section 4 photoelectron spectroscopy as well as how this technique is used to measure Schottky barrier heights will be schematically described; in section 5 recent photoemission results relevant to metal/III-V semiconductor systems will be reviewed; in section 6 the occurence of surface photovoltage phenomena during the photoemission process will be considered   and comparison between the different theories and the experimental data will be discussed; our conclusions will be presented in section 7.

2. INTRODUCTION

An observation of rectifying properties of a metal-semiconductor junction was first reported by F. Braun in 1874 [1]. He observed a deviation from Ohm's law when a voltage was applied across the junction: the resistance varied up to 30 %, depending on direction, intensity and duration of the current. Although the rectification mechanism was not understood, those devices were soon to become widely used in the growing field of broadcasting. It was only in 1938 that Schottky [2] and, independently, Mott [3] succeeded to explain for the first time the general properties of a metal-semiconductor junction by postulating the existence of a potential barrier Φ_B (the so called "Schottky Barrier") at the interface due to the difference in work function between the metal and the semiconductor. Although this model must be considered a mile stone for the understanding of electronic properties of the Schottky barrier (SB), it is not able to predict the SB height value, since it neglects interface effects on electronic properties of the junction. In 1947 Bardeen [4] first proposed the importance of intrinsic surface states in determining the SB height. As a generalization of the Bardeen theory, interface states like defects [5] or metal-induced gap states (MIGS) [6, 7 and 8] were suggested as possible driving forces governing the formation of Schottky barriers. Many scientists now believe that pinning of the Fermi Level is primarily due to MIGS, although a complete agreement between all the experimental data and the theory of Schottky barrier formation is still lacking.

A great deal of experimental studies have been performed to understand the formation of Schottky barriers. Different techniques have been used to gain information, not only on the electronic properties of the junction seen as a technological device, but also on microscopic phenomena occurring at the interface. In this respect, photoemission is one of the most powerful tools to study the chemistry and electronic properties of a metal-semiconductor interface. It allows to follow the Schottky barrier formation as a function of metal coverage, substrate temperature and doping. This is also due to recent developments of technology. Ultra high vacuum is necessary to perform photoemission studies on clean and well characterized samples. In addition, the possibility of using tunable and high intensity synchrotron

radiation, combined with   high resolution monochromators, is of
fundamental importance to perform high-quality experiments.
Using photoemission, the main recent experimental achievements in this
field have been reached, so that in this review only synchrotron
radiation photoemission spectroscopy studies will be discussed. More
detailed information on other experimental techniques is available in
more extensive reviews [9, 10 and 11].

3. THEORETICAL MODELS

Due to the technological importance of metal-semiconductor junctions
much effort has been devoted to produce a theory able not only to
identify the driving forces governing the Schottky barrier formation but
also to predict, with a very high accuracy, the Schottky barrier height
itself. Here some of the most representstive theories proposed in the
past 50 years will be reported. There is still no general agreement on
which of this theoretical models is best capable of explaining all the
experimental findings.

3.1 THE SCHOTTKY MODEL

The basic principles of the Schottky model are shown in fig. 1 in the
case of an n-type semiconductor with a work function smaller than that
of the metal. Fig.1 (a) represents both materials as isolated and
separated from one another. In fig.1 (b) they are brought in electrical
contact so that the two Fermi levels line up. This implies that electrons
pass from the semiconductor to the metal. Thus a negative charge on the
metal surface will be balanced by a positive one in the semiconductor
where an electron-depleted region of width w will be formed. As a
result, the semiconductor bands will bend upwards (the so called "band
bending region" shown in fig. 1). As the metal and the semiconductor
approach one other, the difference V_i between the electrostatic
potentials outside the two surfaces decrease progressively, as shown in
fig.1 (c), until, at intimate contact, it vanishes. In this situation, shown in
fig.1 (d), the Schottky barrier height Φ_{bn} (defined as the energy
difference between the Fermi level and the conduction band minimum),
is given by

$$\Phi_{bn} = \Phi_m - \chi_s \qquad (1),$$

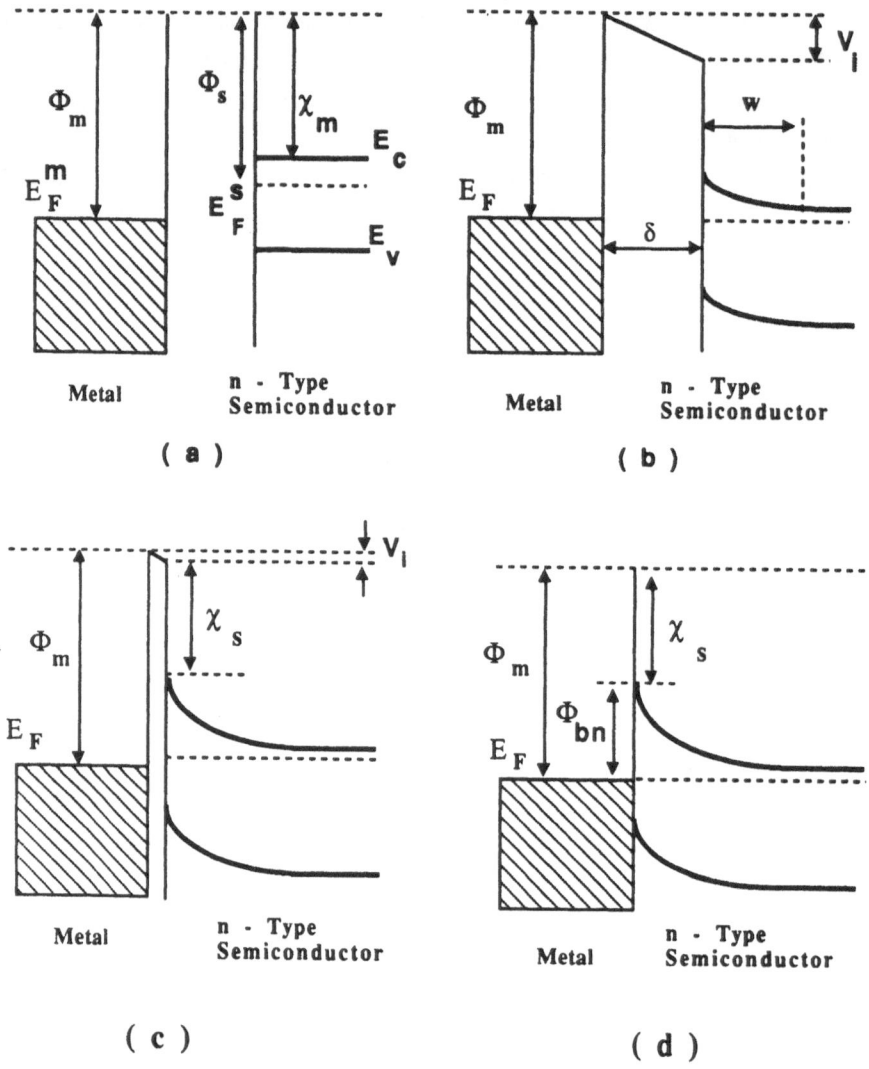

Fig. 1: Formation of a Schottky barrier between a metal and an n-type semiconductor: (a) isolated, (b) electrically connected, (c) separated by a narrow gap, (d) in perfect contact.

whereas in the analogous case of a p-type semiconductor Φ_{bp} (defined as the energy difference between the Fermi level and the valence band maximum), is given by

$$\Phi_{bp} = E_g - (\Phi_m - \chi_s)$$

$$(2),$$

where Φ_m is the metal work function, χ_s is the semiconductor electron affinity and E_g is the semiconductor band gap. This model applies provided that: i) dipole contributions to the Schottky barrier may be neglected; ii) localized states at the semiconductor surface are absent; iii) the junction is atomically abrupt.

In most cases this model failed to predict the correct value of the Schottky barrier height. If we consider junctions formed between the same semiconductor and different metals, the following equation should hold

$$\Delta \Phi_b = \quad s \quad \Delta \Phi_m \qquad\qquad (3),$$

with s = 1 in the Schottky model. It has been determined experimentally that, in most cases, s=0.1 [9, 10, 11]. This implies that the SB height does not depend on the metal work function and that the basic assumptions of this model are to be reconsidered. The only system where an "ideal" Schottky-like behavior was reported [12] (namely metals on GaP(110)) will be discussed in detail in the following.

3.2 THE BARDEEN MODEL

The experimental observations confirming the failure of the Schottky model prompted Bardeen to consider the importance of surface states in pinning the Fermi level. This is shown schematically in fig. 2. Such states, lying at a well defined energy position E_s with respect to the bottom of the conduction band, will be partially filled thus pinning the Fermi level of the clean surface. When the semiconductor is in contact with the metal, a charge flow between the metal and those surface states will occur, until a dipole potential Δ will be formed at the interface. This will compensate for the difference between Φ_m and χ_s. In this model, then, we have

$$\Phi_{bn} = \quad E_s \qquad\qquad (4),$$

so that Schottky barrier heights will not depend on the metal work function. Bardeen stated that his model would work provided that the metal and the semiconductor will remain separated by an insulating layer. Indeed, considering the two materials at intimate contact, the surface states, if any, would be prevented from acting as pinning states by the screening effects of the metal.

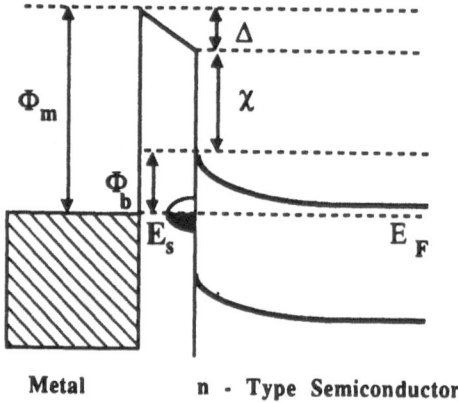

Fig. 2: Metal-semiconductor contact with surface states.

This theory's shortcoming can be solved by considering interface states and not surface states as pinning the Fermi level. More recent theories, to be described shortly, are based on this approach with a view to identifying the origin of these interface states.

3.3 THE DEFECT MODEL

In 1979 Spicer et al. [5] proposed a model for metal/III-V semiconductor junctions. Metal deposition will induce defects characteristic to the semiconductor, with a definite energy E_d (in the case of donor defects pinning a n-type semiconductor) and E_a (in the case of acceptor defects pinning a p-type semiconductor). Those native defects can extend away from the surface, thus they will not be necessarily screened by the metal at intimate contact. In such a case, if E_d is referenced to the conduction band minimum and E_a to the valence band maximum, we have

$$\Phi_{bn} = E_d \text{ (n-type semiconductor)}$$
$$\text{and} \qquad (5)$$
$$\Phi_{bp} = E_a \text{ (p-type semiconductor)}$$

E_d and E_a will not have in general the same position within the semiconductor energy gap so that $\Phi_{bn} + \Phi_{bp} \neq E_g$. This implies that the final Fermi level pinning position will be different for a junction formed between the same metal and p- or n-doped semiconductor. In most cases, this difference cannot be observed experimentally. This fact reduces the validity of this model. Nevertheless, up to now, one cannot exclude the existence of these defect states, and their importance in pinning the Fermi level, at least in some particular cases such as at submonolayer coverages.

3.4 THE "MIGS" MODEL

A different approach is based on a model proposed by Heine [6] .
He pointed out that when the metal conduction band overlaps with the semiconductor energy gap, states will be formed from the tailing of the metal wave functions into the semiconductor gap. Those states are called "Metal Induced Gap States" (MIGS) and they match the bulk semiconductor states both in phase and in amplitude. An example of this can be seen in fig. 3, where theoretical calculations by Louie and Cohen [13] of the local density of states at the Al/Si(111) interface.are reported. In the interface region, on the semiconductor side, MIGS are clearly visible. The Schottky barrier height will then strongly depend on these states which will pin the Fermi level within the gap. An extension of this model was presented by Tersoff [7] in 1984. He noticed that, since the bulk semiconductor band structure is formed by a complete set of wave functions, the induced gap states cannot be considered as additional states, but, rather, as a linear combination of the semiconductor valence and conduction states. There is an energy E_n (called the "charge neutrality level"), within the semiconductor gap, where the MIGS wave functions cross over from being largely valence-band derived to being conduction band derived. Then the Fermi level is pinned at E_n for both n- and p-doped semiconductors, or, in a more general case, the Schottky barrier is given by

$$\Phi_b = E_n + \frac{(\Phi_m - \chi_s - E_n)}{k} \qquad (5),$$

where k is a screening factor related to the semiconductor dielectric function.

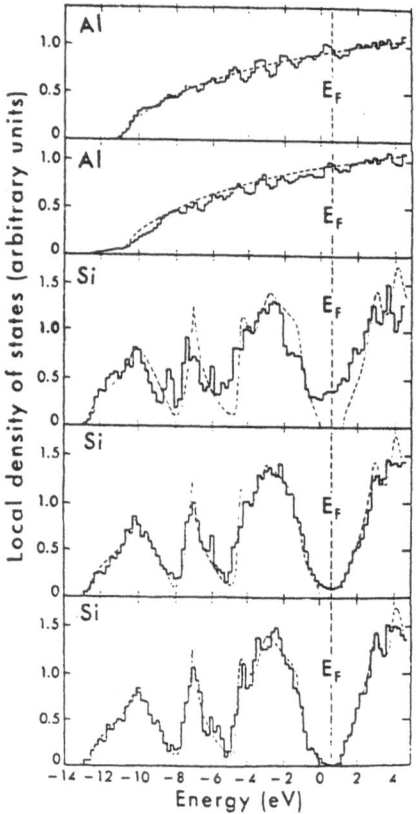

Fig. 3: Local density of states on either side of the junction for Al on Si(111) (from
   ref. [13]).

In the one-dimensional case, it can be easily shown that E_n is exactly at midgap. In the three-dimensional case of interest, E_n can be calculated from semiconductor bulk properties. Indeed, Tersoff's predictions of Schottky barrier heights differ from the experimental data by about ± 0.2 eV. It is difficult to explain with this theory the Schottky barrier behavior at least in the submonolayer metal coverage regime. Some recent low-temperature photoemission experiments [14 and 15]   have shown that a strong correlation exists between the onset of the overlayer metallicity and the reaching of the final Schottky barrier height value. This evidence is now considered to be the final proof supporting MIGS model.

4. USE OF PHOTOEMISSION TO STUDY SCHOTTKY BARRIERS

Synchrotron radiation photoelectron spectroscopy is regarded as one of the best techniques to investigate Schottky barrier formation. Indeed, even if electrical measurements such as I-V characteristics or the capacitance-voltage method [9] can give the Schottky barrier height Φ_b more accurately than photoemission, these methods fail to give complementary information on the microscopic properties of the junction. This information has been shown to be a fundamental issue to gain a complete insight of what really happens at the interface.

After a brief description of photoemission techniques, it will be shown how to extract the Schottky barrier height from a "band bending experiment", as well as how to get chemical information by means of core level line shape analysis.

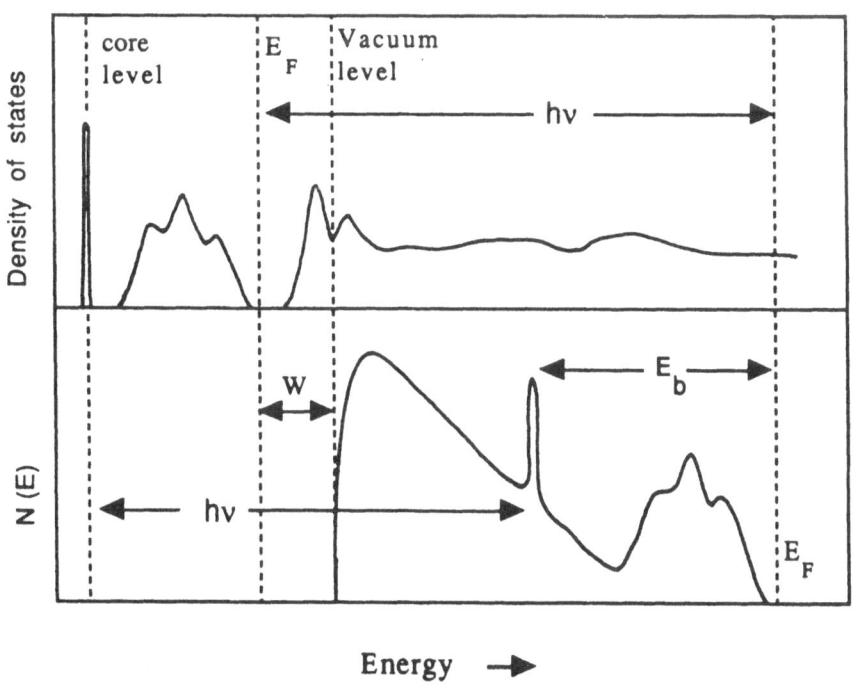

Fig. 4:   Schematic diagram of photoelectron spectroscopy.

4.1   PHOTOEMISSION

Photoelectron spectroscopy has been the subject of careful investigations during the last years, and a great deal of literature on the topic is available [16, 17, 18 and 19]. Here, only a brief outline is given on those aspects relevant to the subject of this paper.

The basic physics of the photoemission process is schematically shown in fig. 4: when photons of energy hv impinge on a surface, some of them are absorbed transfering their energy to core or valence electrons. Some of these excited photoelectrons are emitted from the sample and can be detected. An Energy Distribution Curve (EDC) (fig.4) consists of features related to electrons which did not suffer any energy loss before emission as well as to those which has scattered inelastically (secondary electrons). Analysis of the EDC's as a function of spin, emission angle and kinetic energy gives information on the electronic states (spin, momentum and binding energy) within the sample. The difference between E_k and E_F governs the inelastic mean free path for electrons in solids which (as seen in   fig.5) is rather short - typically of the order of two atomic layers for electrons with kinetic energy in the range 20 to 100 eV - and so only primary photoelectrons created near the surface can be detected. This makes photoemission from solids an intrinsically surface-sensitive technique, ideal to study the properties of a forming Schottky barrier.

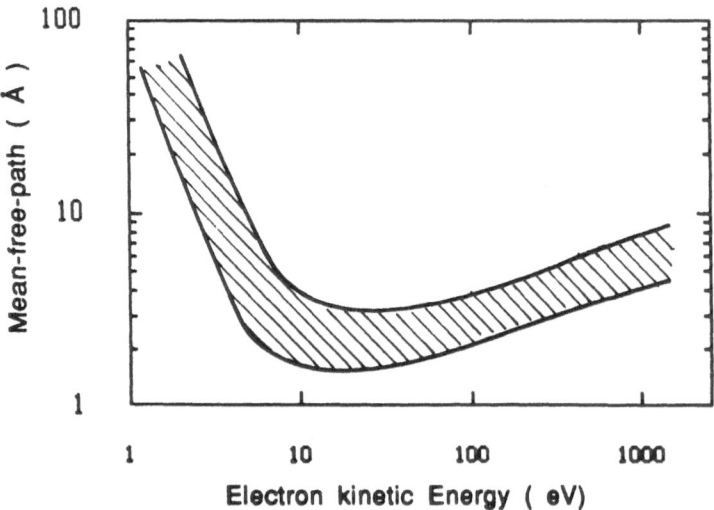

Fig. 5: Mean free path range of electrons in solids as a function of energy.

4.2. SCHOTTKY BARRIER HEIGHT MEASUREMENTS

A band bending experiment is basically a simple experiment: core levels and valence band energy distribution curves are first measured for the clean surface, then the EDC series is repeated while increasing metal coverages are deposited step by step onto the surface. Since in most photoemission spectrometers binding energy is referenced to the Fermi level of the sample, the introduction of band bending due to the deposited metal, will cause a rigid shift of measured binding energies. Photoemission sampling depth d (see fig.6 a) is of the order of 10 Å and the depletion region w   (see fig.6 b) in a normally doped semiconductor $(10^{17} - 10^{18}$ atoms per $cm^3)$ is of the order of 1000 Å . Hence, only electrons from the topmost part of the depletion region will be detected.

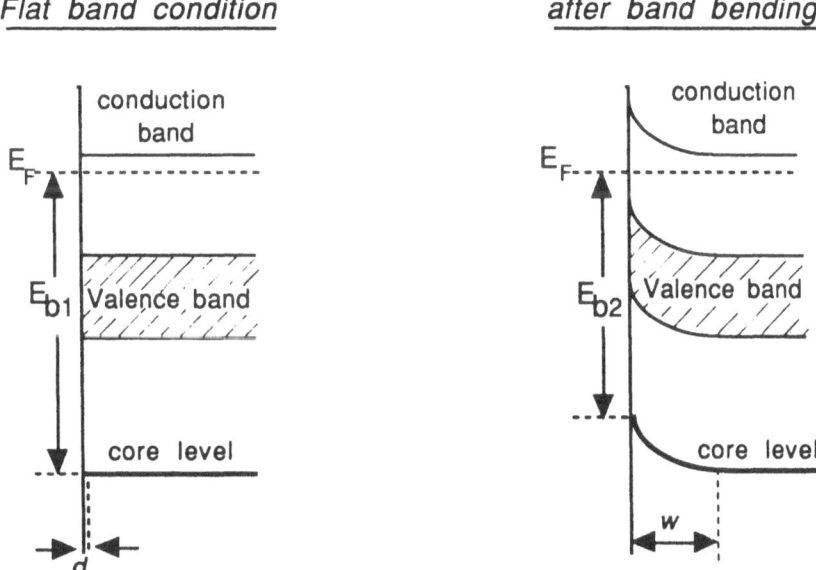

Fig. 6: Illustrating how the presence of band bending is reflected by a change in binding energy of a core level at the surface with respect to the Fermi level.

This situation is shown in fig. 6 (left panel) where it is assumed that the clean surface is in the so called "flat band" condition occurring when, in absence of any surface pinning states, the position of the surface Fermi level, relative to the band gap edges, is the same as in the bulk semiconductor and is a function of bulk properties alone, like the doping type and quantity. The situation after metal coverage has induced some band bending is shown in fig.6 (right panel). If $w >> d$, the difference in the core level binding energy $E_{b1} - E_{b2}$ is equal to the band bending contribution induced by that metal coverage. In general, in photoemission studies of Schottky barrier, the semiconductor gap edges are assumed as energy reference; then band bending changes will be reflected by a shift in the surface Fermi level position within the semiconductor gap. Plots of this quantity, extracted by measuring rigid shifts in the substrate core level positions, as a function of metal coverage (see fig. 7 as an example) are generally presented to discuss the formation of a Schottky barriers.

4.3 CORE LEVEL SHIFTS

The measured binding energy of a substrate core level can change after metal deposition not only because of band bending, but also because of simultaneous chemical shifts produced by the bonding between the metal and the semiconductor. This makes it sometimes difficult to extract the accurate value of the rigid shift due to band bending, so that the use of high resolution core level spectroscopy is needed to assure validity to band bending studies.
The origin and characteristics of core level chemical shifts have been analyzed in detail in the literature [19]. A detailed discussion is beyond the scope of this review. Nevertheless, it is important to point out that core level photoemission spectroscopy has been widely used to characterize a metal-semiconductor interface, since core level analysis can yield information on the mode of adsorption of the adsorbate, on its reactivity with the substrate, and on its growing mode.

5. PHOTOEMISSION STUDIES ON SCHOTTKY BARRIERS

Since early photoemission studies on Schottky barriers, it was clear that, in most cases, the final barrier height was mainly independent of the nature of the metal deposited and that this barrier was already formed after deposition of only a metal monolayer. This indicates that the driving forces governing the Schottky barrier height should be operating at coverages of a monolayer and below. Hence interest arises in studying metal-semiconductor interfaces in the so called *"ultra low coverage regime"* namely depositing metal coverages ranging from 10^{-3} to - 2 ML. The absence of surface states within the gap of a III-V semiconductor clean (110) surface [9], is the reason why most of these studies are performed on such substrates. In the following I will concentrate only on these studies, leaving the description of other systems to more detailed reviews [9-11].

Furthermore, additional studies have been performed at low temperature (liquid nitrogen or below) in an attempt to single out different possible mechanisms that can be simultaneously acting at room temperature. If interface reactivity is not reduced by decreasing the temperature [20], overlayer morphology is generally changed. Many metals are expected to form clusters and islands after deposition at RT. At low temperature, on the other hand, a more laminar growth is expected [8]. Those experiments, although of general relevance for metal on III-V semiconductor, have been performed mainly on GaAs(110) and will be described in the next section.

5.1 METAL ON GaAs(110) : RECENT RESULTS

A general situation has been observed when studying different metals on GaAs(110) as a function of metal coverage, semiconductor doping type, and temperature. This is schematically shown in fig.7: at room temperature the surface band bending develops quite symmetrically on p- and n-doped samples, while at low temperature a strongly asymmetric behavior is observed [8]. The analysis of such a behaviour reported in the literature [8-15] suggests the importance of two distinct types of electronic states in determining the surface position of the Fermi level within the semiconductor band gap. When metallic islands or continuous metallic films are formed MIGS states pin the Fermi level to its final position equal in both temperature regimes and doping types. At low temperature, in absence of clustering, adatom-related surface states of donor character have been proposed [8] as being responsible for the observation reported in fig.7: on a n-type semiconductor, no change in the surface Fermi level position occurs at LT until overlayer "metallicity" is reached; on the contrary, on a p-type semiconductor, a pinning position, generally different from the final one (and higher), is reached at very low coverages, then the final SB height is reached at overlayer metallicity. This intermediate plateau is supposed to reflect the energy position of donor states in the semiconductor gap. At room temperature and very low coverages, both mechanisms can be effective according to the growth morphology and interface chemistry [8] but also the energy of charging microscopic metal clusters have been indicated as a possible driving force, pinning the Fermi level [21].

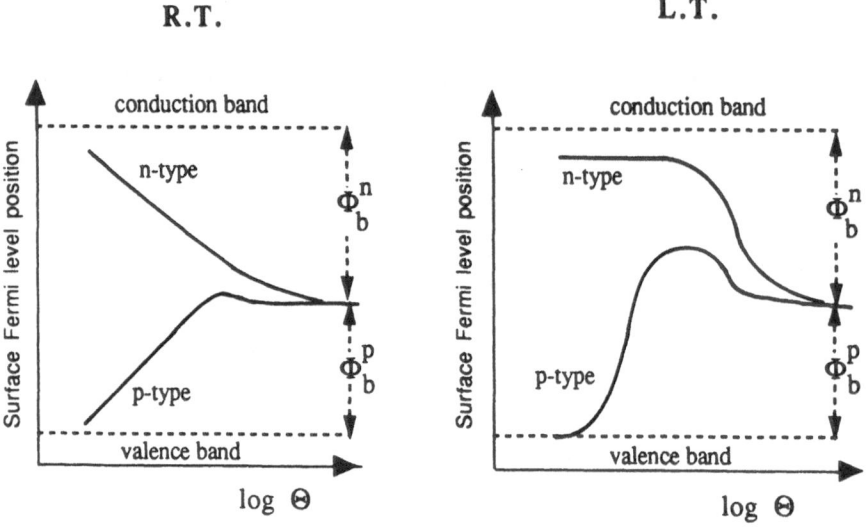

Fig. 7: General behavior of the surface Fermi level position within the semiconductor gap, observed for different metals on GaAs(110), as a function of metal coverage (logΘ), semiconductor doping type, and temperature.

Recently Weaver's group reported experimental results showing that the position of the surface Fermi level for GaAs(110) depends critically on the bulk dopant concentration N and the temperature T at which the measurements are made, when adatoms of a wide variety of metals are deposited [22, 23, 24 and 25]. This is shown in Fig. 8 in the case of Ti/GaAs(110). At low doping ($\sim 10^{17}$ cm^{-3}) the coverage dependent surface Fermi level position stays near the gap extrema for both p- and n-type until the onset of metallicity is reached. It then moves symmetrically into the gap. For higher dopant levels ($\sim 2 \times 10^{18}$ cm^{-3}), E_F moves before metallicity and the step is reduced or lost altogether, as shown in fig. 8 a). Temperature dependent studies for T ranging from 20 to 300 K at a fixed coverage Θ, demonstrate that, in absence of morphological changes, E_F can be moved reversibly into the gap, as seen in fig.8 b). A dynamic coupling between adatom induced states and substrate states has been proposed by the same authors to explain this unexpected dependence of the surface Fermi level position on the bulk semiconductor doping and temperature.

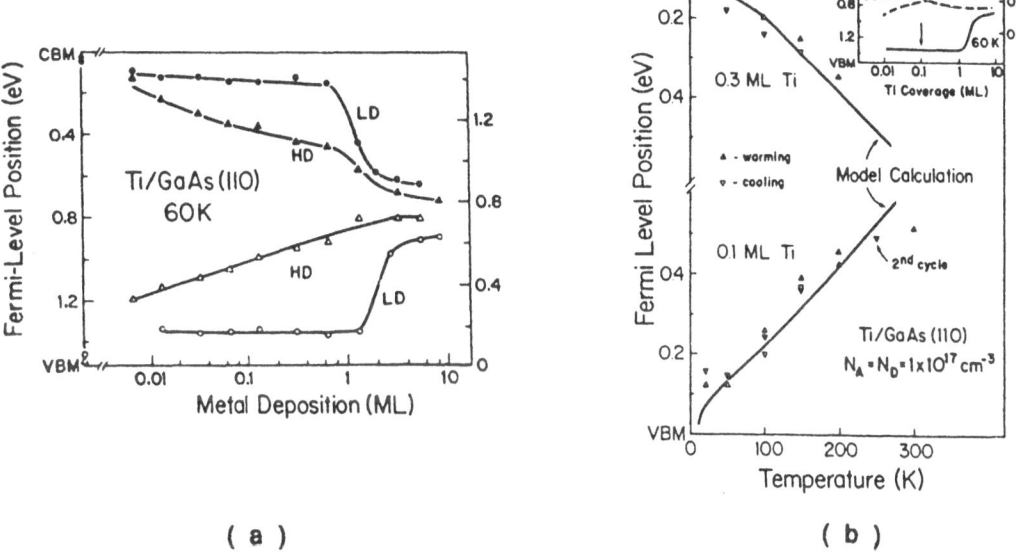

(a)   (b)

Fig. 8 : Fermi level position for Ti/GaAs(110) system: a) at 60 K as a function of semiconductor doping and metal coverage; b) as a function of temperature (ref. [22-24])

This model suggests a fundamentally non-chemical character of Schottky barrier formation in striking contrast with the existing models presented in section 3.

Experimental studies of interfaces formed between different metals and GaP (110) by Alonso, Cimino and Horn [26, 27], along with a contemporary theoretical work by Hecht [28, 29] have suggested that Weaver's group observations and, more generally, all the experiments performed at low temperature, can be affected by surface photovoltage

effects. The consequences of such phenomena will be discussed in detail
in the next section.

5.2 METAL ON GaP(110) : RECENT RESULTS

Photoemission results for various metal/GaP(110) junctions seem to
indicate that these systems reveal an almost ideal Schottky-like
behavior, as shown in fig. 9 [12].

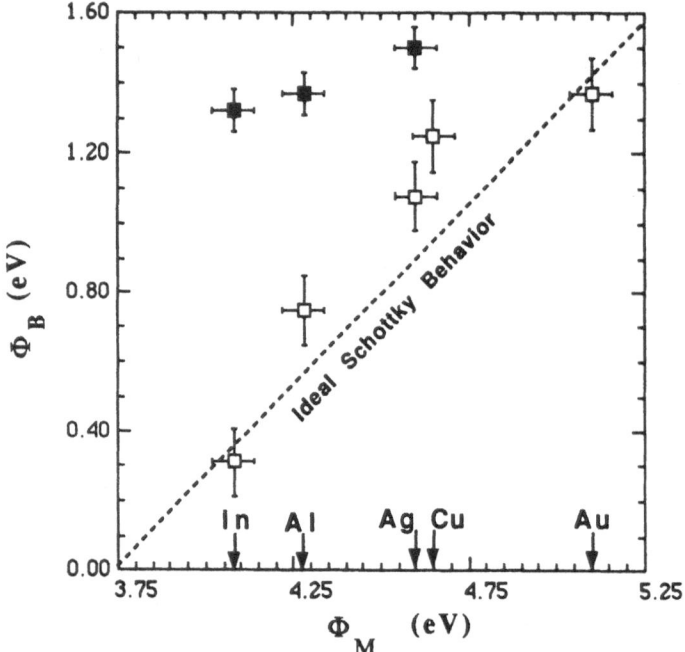

Fig.9: Schottky barrier values Φ_b vs metal work function minus semiconductor
electron affinity ($\Phi_m - \chi$). Open symbols from Ref. [12]; full symbols
from Ref. [27]; the dotted line shows the values according to the
Schottky-Mott theory (see Sec. 3.1).

This is in contrast with the metal/GaAs case, where no dependence on
the metal work function is found for the Schottky barrier height. A more
careful study of metals on GaP is then a very important issue because it
may reveal possible limits of the theory used to understand the driving
forces governing the Schottky barrier formation on GaAs(110). This
problem has been addressed recently by other authors [26, 27]. They

found no dependence on metal work function for the final Schottky barrier height for the different system studied (full dots in fig. 9). Although the reason of the discrepancy between the two sets of data is not fully understood at present, it is clear from Ref. [26] that surface photovoltage phenomena can heavily affect band bending studies of metals on GaP(110), both at room (RT) and low temperature (LT), and hence the determination of Schottky barrier final heights.

6. SURFACE PHOTOVOLTAGE PHENOMENA

The photovoltaic effect, i.e. the appearance of a voltage upon illuminating a semiconductor-metal rectifier, was first reported more than one hundred years ago [30]. Since 1948 this effect has been used to study semiconductor surfaces [31,32] and a spectroscopy is based on this very effect [33,34]. The basic principle of this phenomenon is simple and is schematically shown in fig. 10 in the case of a metal overlayer on an n-type semiconductor.

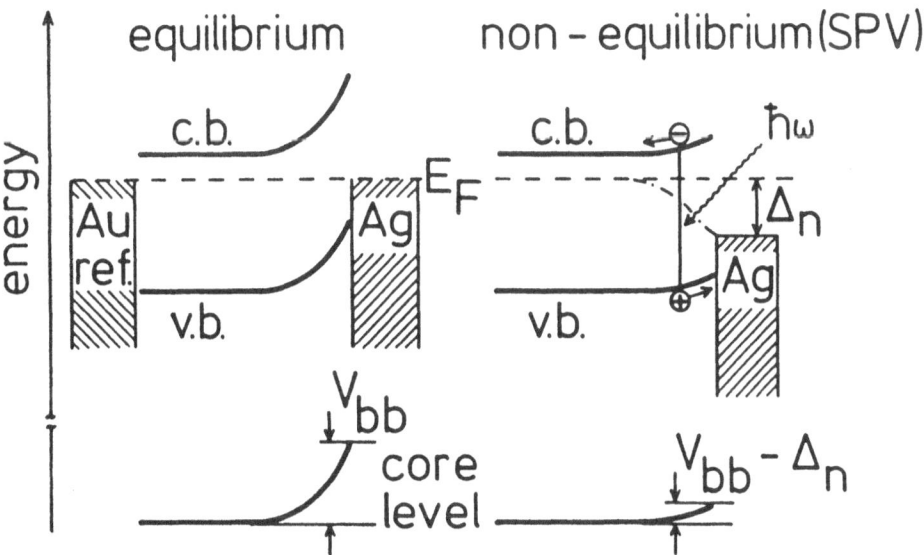

Fig. 10: a) schematic picture of the equilibrium level alignment;   b) the effect of
surface photovoltage on the energies of the substrate and the overlayer.

In the equilibrium situation of fig. 10 a) the presence of the metal overlayer gives rise to a surface band bending V_{bb}. Then, electron-hole pairs created by the incident light and/or by secondary processes can be separated, if recombination can be neglected, by the built-in electric field present in the semiconductor depletion region. This will create the non-equilibrium situation, as shown in Fig.10 b). Electrons are driven into the semiconductor bulk whereas holes are trapped near the surface, compensating for the space charge and shifting rigidly all the semiconductor and metal levels. Recent theoretical work by Hecht [28,29] predicts that surface photovoltage (SPV) can strongly affect the experimental results obtained at low temperature. The calculated SPV strength versus semiconductor doping type and temperature explains quite accurately the experimental data by Weaver's group described in section **5.1**. Without going into details on how surface photovoltage can be calculated [32, 33, 28, 29], it is clear that, depending on the recombination efficiency of the electron-hole pairs, SPV will be more efficient for low-doping, large gap semiconductors and at low temperature. Secondary illumination effects on photoemission spectra from semiconductor surfaces have been reported in 1980 by Margaritondo et al. [35], whereas the presence of surface photovoltage induced by uv irradiation (that is, intrinsic to the photoemission process) has been successfully used by Demuth et al. [36] and, more recently, by Myler and Jacobi [37] to measure Schottky barrier heights of different clean silicon surfaces. Both groups noted that a saturation surface photovoltage occurs on these systems at substrate temperatures lower than 50 K. This SPV gives rise to "flat band emission", so that a temperature dependent photoemission study provides a simple and direct method for determining band bending and barrier heights.

Although these effects may occur on clean as well as metal-covered semiconductor surfaces [36-38], the possibility that the photoemission data could be affected by photon-induced electron-hole pairs creation and transport processes leading to a nonequilibrium charge distribution has been widely overlooked in photoemission studies of Schottky barriers even if performed at low temperature.

Experimental observations on the effect of surface photovoltage on photoemission studies of Schottky barriers and the consequences of this phenomenon on previous works on Schottky barriers, will be critically discussed in the following section.

6.1 SURFACE PHOTOVOLTAGE EFFECTS ON PHOTOEMISSION FROM METAL-GaP(110) INTERFACES

Surface photovoltage effects on Schottky barrier's photoemission experiments have been first observed in a set of studies where aluminium, indium and silver have been deposited on GaP(110) as a function of metal coverage, substrate temperature and doping. These metals have been chosen for their different reactivity with the other III-V semiconductors [38].

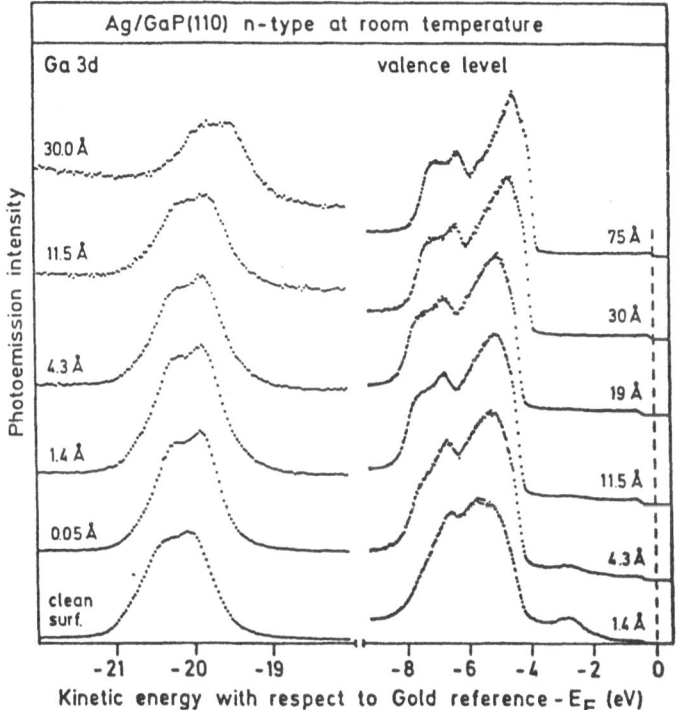

Fig. 11:   Sets of spectra for different doses of Ag on n-type GaP(110) deposited at room temperature showing the Ga 3d and the valence band regions. The spectra, obtained at hv=60.0 eV, are normalized to the same height to emphasize line shape changes.

As confirmed by a high resolution photoemission core level study performed on those systems [27] Al is highly reactive, whereas both In an Ag can be considered as examples of unreactive metals. Temperature reduction causes a more laminar growth in all of the cases studied. This strong difference in reactivity and the morphological changes at different temperatures may have an effect on the coverage dependence of the surface Fermi level, at least for thin overlayers [39]. Therefore this study should give some new insight on the formation of metal-GaP(110) interfaces.

In fig.11 a typical series of selected Ga 3d core level and valence band spectra is given for increasing Ag coverage on n-type GaP(110) at room temperature. The Ga 3d peak shows very little changes, except for some broadening and filling of the small valley between the two spin-orbit components at around 11.5 Å deposition and beyond. This confirms the very limited interface reactivity of this system [40].

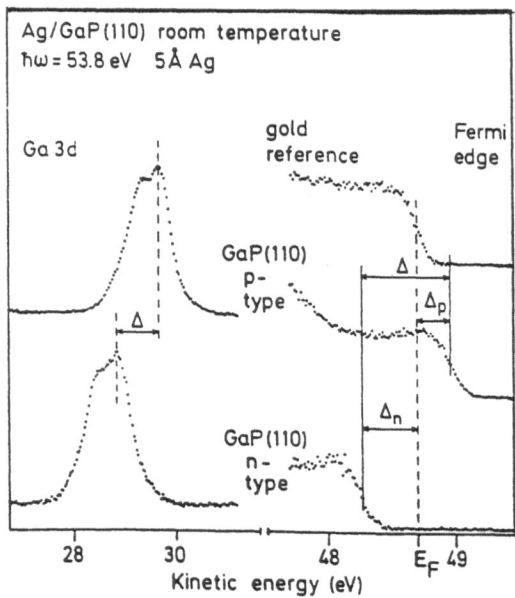

Fig. 12:   Influence of SPV on the semiconductor core level and metal Fermi level of a 5 Å   Ag layer on n- and p-type GaP(110) deposited at room temperature. Note the different energy scales for core and valence level spectra. Also shown is the reference Fermi level from an Au foil in contact with the samples.

In the valence band region the GaP features are still present up to 4.3 Å, then the stronger emission of the Ag 4d becomes dominant, evolving into sharper structures with increasing coverage.   Measuring Ga core level shifts as a function of metal coverage is meant to give us the band bending evolution of this interface. The fact that photon-induced nonequilibrium phenomena affect the determination of surface band bending at Ag/GaP(110) is more evident from fig.12, where valence band maxima and Ga 3d core levels are reported in the case of 5 Å Ag evaporation onto n- and p-type GaP(110).

At this coverage the Ag Fermi level becomes detectable. For direct comparison the reference Fermi level from an Au foil in thermal and electrical contact with the samples is shown. Unusual occurence of photoelectrons at a kinetic energy above the reference Fermi level for p-type substrates (shifted by Δ_p) is evident in fig 12, while on the n-type substrates the Ag Fermi level emission occurs below this level (shifted by Δ_n). Also the total distance between the two core levels is coincident, within 100 meV, with the total shift $\Delta = \Delta_p + \Delta_n$ observed for the top of the valence bands. This indicates that the shift is different in sign for the two doping types and equally affects overlayer and substrate emission features. Charging can be excluded as a source of the shifts, as checked in experiments on the clean surface; it is also incompatible with the observed shifts because of the opposite sign of Δ_p and Δ_n. This fact also rules out charging of metal clusters as observed for Ag on graphite by Wertheim et al. [41]. As described in the previous section, these observations can be readily explained by the occurrence of surface photovoltage caused by the incident light used for photoexcitation. The study of such effect as a function of coverage and temperature gives some important hints on the dependence of SPV on different parameters. Selected spectra are shown in fig. 13. At RT the clean surface valence band maximum (VBM) of n-type GaP(110) has the Fermi level pinned at about 1.8 eV above VBM, most likely due to empty surface states in the gap of GaP(110) [42]. Lowering the temperature gives nearly "flat band" conditions due to the strong enhancement of the SPV strength at LT. As soon as the Ag Fermi level emission is detectable, the effect of SPV can be measured, and the data can be therefore corrected. At higher coverages, the Ag Fermi level emission shifts progressively towards the reference level E_F, finally coinciding with it at a dose of 110 Å Ag. This is probably due to the fact that since Ag grows in clusters, at high coverage a conduction path is formed on the surface shorting the SPV still present to ground.

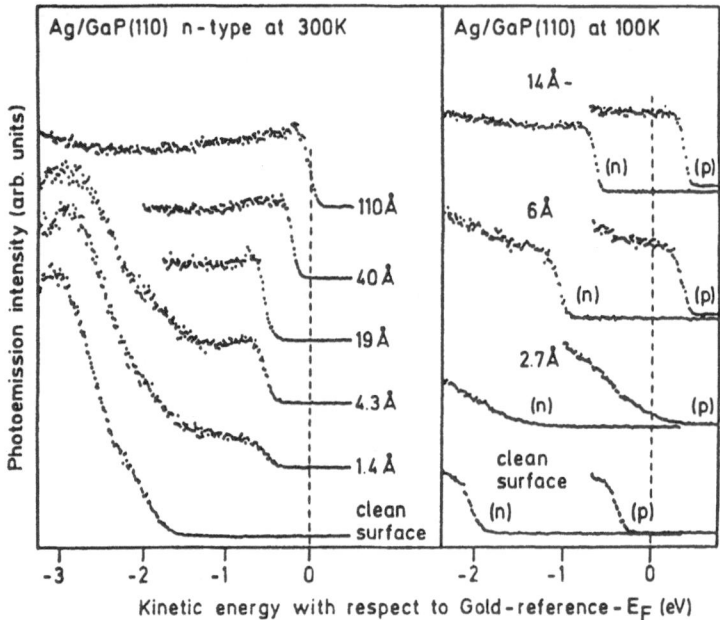

Fig. 13:   Spectra of the valence band edge of GaP(110) with different depositions of Ag, at room temperature(left-hand side) and low (100 K) temperature (right-hand side), showing the magnitude of the SPV-induced shifts as a function of Ag overlayer thickness (hv=60.0 eV).

This effect makes actually difficult to compare RT with LT data since, if SPV is stronger at LT, due to a more laminar overlayer growth, surface conductivity will be more efficient in reducing SPV to zero. This is confirmed by the observation that the progressive shift of the Fermi edge towards the gold reference value is observed at much lower nominal coverages for LT than for RT.

The large SPV observed here in photoelectron spectroscopy from the different systems studied and its persistence up to high metal depositions, has strong implications upon the possibility of a correct determination of temperature and coverage-dependent surface band bending and Schottky barrier height by photoemission. This is shown as a function of metal deposition in fig.14 for the case of In on n- and p-type GaP(110) substrates at liquid Nitrogen temperature(circa 100 K).

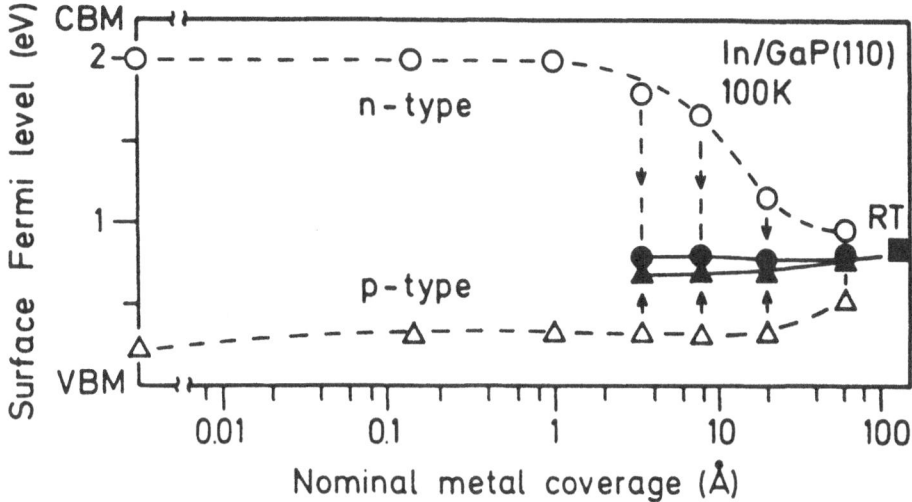

Fig. 14:   Position of the surface Fermi level in the semiconductor band gap, as a
             function of nominal metal coverage. Open symbols give the
             uncorrected values, obtained by the method normally presented in the
             literature, whereas full symbols correspond to.the real Fermi level
             position, i.e. after correction for surface photovoltage.

Open symbols give the values as extracted from the core level shifts,
whereas solid symbols show the position after correcting by SPV as
measured for the two different doping types by the quantities Δ_p and
Δ_n, respectively. Note the high similarity between this diagram without
SPV correction and the one reported in fig. 8 from the work of Aldao et
al. [25] : nearly flat bands for n- and p-type substrates at low metal
coverages, and nearly a symmetric behaviour for n- and p- materials.
Note also, at higher coverages, the drop from flat bands towards midgap
observed on the uncorrected n-type curve; it exhibits a close
resemblance to the shape of the curve typically found for Schottky
barriers formed at low temperatures on n-type III-V semiconductors
(see fig 7). Hence, it seems that most of the conclusions from the data
shown here for metal/GaP(110) systems can reasonably be generalized
to other metal/III-V(110) interfaces formed at low temperature, and in
particular, to those grown on low doped GaAs(110).
Keeping in mind such considerations let us examine the values shown in
fig.14 after the SPV correction (solid symbols). The situation is quite
different; the E_F positions from n- and p-type substrates, which

apparently were widely separated (1eV) before the SPV correction, almost coincide now (within 100 meV). The drop in E_F typically observed at LT on n-type substrates, which accompanies the appearance of metallicity at the interface and was often considered [8, 14, 15] as an experimental evidence for the MIGS model, disappears after SPV correction. Even considering the difficulty of measuring SPV at low metal coverages, the data presented here clearly reveal that the Fermi level is pinned in a single position for n-and p-type samples before the observed drop occurs. The drop, as well as the previous flat band conditions, are simply an artifact of the measurement technique. Experiments on GaP(110) surface with other metals (reactive and non-reactive) confirm the generality of these conclusions, which are also in agreement with other results by Cimino et al. on In/GaAs(110) [42], by Waddill et al. on Bi on low doped GaAs(110) [43], by Chang et al. on clean GaAs(100) surfaces studied by photoemission as a function of temperature [44], by Mao et al. for Ag/GaAs(110) studied by combining synchrotron radiation with Kelvin probe measurements [45].

7. CONCLUSIONS

It has been shown that, in general, the underlying assumption of most of the previous photoemission work on band bending and Schottky barrier height determination, that the photoelectron spectra reflect the equilibrium valence band arrangement, needs experimental confirmation in each single case. Moreover, the possibility of measuring directly the SPV strength only by determination of the absolute energy position of the surface Fermi level from the emission of the metallic overlayer implies a reduction in the capability of photoemission to extract reliable information on the formation of Schottky barriers. In fact, only in some cases, and for a nearly metallic overlayer, is it possible to clearly demonstrate the nonexistence of SPV or, if present, to measure its strength. This makes results obtained at ultra low coverages (from 0. up to circa 1/2 ML) quite ambiguous.

This does not mean that synchrotron radiation photoelectron spectroscopy did not and will not give a valuable help in understanding the problem of Schottky barrier formation. It means that a more careful attitude in the analysis of photoemission spectra, together with some support from other techniques like Kelvin probe [45], scanning tunneling microscopy (STM) [46] and ballistic electron emission microscopy (BEEM) [47] is needed to gain information on the driving forces governing the electronic properties at metal-semiconductor interfaces.

8. ACKNOWLEDGMENTS

The author would like to thank M. Alonso and K. Horn for unvaluable discussions on the physics of Schottky barriers as well as for creating the stimulating atmosphere that made possible a fruitful collaboration on SPV studies. The author also thanks F. Boscherini, A. Giarante, E. Paparazzo and M. Pedio for reading the manuscript.

REFERENCES

* Present address: INFN-LNF P. O. Box, 13, I-00044 Frascati (Italy).

[1] F. Braun, *Pogg. Ann.* **153**, 556 (1874).

[2] W. Schottky, *Naturwissenschaften* **26**, 843 (1938)

[3] N. F. Mott, *Proc. Cambridge Philos. Soc.* **34**, 568 (1938)

[4] J. Bardeen, *Phys. Rev.* **71**, 717 (1947)

[5] W.E. Spicer, P.W. Chye, P.R. Skeath, C.Y. Su, and I. Lindau; *J.Vac.Sci. Technol.* **16**, 1422 (1979).

[6] V. Heine, *Phys. Rev.* **138**, A 1689 (1965).

[7] J. Tersoff, *Phys. Rev. Lett.* **52**, 465 (1984).

[8] W. Mönch, *J.Vac. Sci. Technol.* B **6**, 1270   (1988)

[9] L.J. Brillson, *Sur. Sci. Rep.* **2**, 123 (1982)

[10] "The Chemical Physics of Solid Surfaces and Heterogeneous Catalysis" edited by D.A. King and D.P. Woodruff; vol. 5, pags. 119-182, Elsevier Science Publisher B. V., 1988.

[11] "Metal-Semiconductor Contacts" edited by E.H. Rhoderick and R.H. Williams, Clarendon press - Oxford- (1988)

[12] P. Chiaradia, L. J. Brillson, M. Slade, R.E. Viturro, D. Kilday, N. Tache, M. Kelly, G. Margaritondo, *J.Vac. Sci. Technol.* B **5**, 1075   (1987); and P. Chiaradia, M. Fanfoni, P. Nataletti, P. DePadova, R.E. Viturro, L. J. Brillson, *J.Vac. Sci. Technol.* B **7**, 195   (1989).

[13] S.G. Louie, M.L. Cohen *Phys. Rev.* B **13**, 2461 (1976).

[14] M. Prietsch, M. Domke, C. Laubschat and G. Kaindl; *Phys. Rev Lett.* **60**, 436 (1988).

[15] K. Stiles and A. Kahn; *Phys. Rev. Lett.* **60**, 440 (1988).

[16] M.Cardona and L. Ley (ed.), 1978 *Photoemission in solids I, II Topics in Applied Physics* vol **26** (Springer-Verlag Berlin)

[17] B. Feuerbacher, B. Fitton and R.F. Willis (ed.) 1978 *Photoemission and Electronic Properties of Surfaces* (Wiley, Chichester).

[18] N.W. Smith and F.J. Himpsel, 1983, in: *Handbook of Synchrotron Radiation,* Vol. **1**, ed. E. -E. Koch (North Holland, Amsterdam) p. 955.

[19] W. F. Egelhoff, Jr. *Surface Science Reports* **6**, 253-415 (1987).

[20] A. B. McLean, I. T. McGovern, C. Stephens, W. G.Wilke, H. Haak, K. Horn, W. Braun *Phys. Rev.* **B 38**, 6330 (1988).

[21] S. Doniach, K. K. Chin, I. Lindau, W. E.Spicer, *Phys. Rev. Lett.* **58**, 591 (1987).

[22] C. M. Aldao, Steven G. Anderson, C. Capasso, G. D. Waddill, I. M. Vitomirov, and J. H. Weaver *Phys. Rev.* **B 39**, 12977 (1989).

[23] I. M. Vitomirov, G. D. Waddill, C. M. Aldao, Steven G. Anderson, C. Capasso, and J. H. Weaver *Phys. Rev.* **B 40**, 3483 (1989).

[24] Steven G. Anderson, C. M. Aldao, G. D. Waddill, I. M. Vitomirov, S. J. Severtson, and J. H. Weaver, *Phys. Rev.* **B 40**, 8305 (1989).

[25] C. M. Aldao, I. M. Vitomirov, G. D. Waddill, Steven G. Anderson, and J. H. Weaver *Phys. Rev.* **B 41**, 2800 (1990).

[26] M. Alonso, R. Cimino, and K. Horn, *Phys. Rev. Lett.* **64**, 1947 (1990).

[27] M. Alonso, R. Cimino, Ch. Maierhofer, Th.Chasse', W.Braun, and K. Horn, *J.Vac. Sci. Technol.* **B 8**, 955 (1990).

[28] M. H. Hecht *Phys. Rev.* **B 41**, 7918(1990).

[29] M. H. Hecht *J.Vac. Sci. Technol.* **B 8**, 1018 (1990).

[30] W. G. Adams and R.E. Day, *Proc. R. Soc.* **A 25**, 113 (1876)

[31] W. H. Brattain, *Phys. Rev.* **72**, 345 (1948).

[32] E. O. Johnson, *Phys. Rev.* **111**, 153 (1958).

[33] Harry C. Gatos and Jacek Lagowsky, *J.Vac. Sci. Technol.* **10**, 130 (1973).

[34] L. J. Brillson *Phys. Rev.* **B 18**, 2431 (1978).

[35] G. Margaritondo, L. J. Brillson, and N. G. Stoffel, *Solid State Comm.* **35**, 277 (1980).

[36] J. E. Demuth, W. J. Thompson, N. J. DiNardo and R. Imbihl, *Phys. Rev. Lett.* **56**, 1408 (1986).

[37] U. Myler and K. Jacobi, *Surface Science* **220**, 353 (1989).

[38] K. Stiles, A. Kahn, D. Kilday and G. Margaritondo; *J. Vac. Sci. Technol.* **B 5**, 987 (1987) and references therein.

[39] W. Mönch, *Europhhysics Lett.* **7**, 275 (1988).

[40] B. M. Trafas, F. Xu, M. Vos, C. M. Aldao and J. H. Weaver, *Phys. Rev.* **B 40**, 4022 (1989).

[41] G. K. Wertheim, S. B. DiCenzo and S. E. Youngquist, *Phys. Rev. Lett.* **51**, 2310 (1983).

[42] R.Cimino, M. Alonso, K. Horn, to be published

[43] G. D. Waddill, Tadahiro Komeda, Y. -N. Yang, and J. H. Weaver *Phys. Rev.* **B 41**, 10283 (1990).

[44] S. Chang, I.M. Vitomirov, L. J. Brillson, D. F. Rioux, P. D. Kirchner, G. D. Petitt, J. M. Woodall, M. H. Hecht, *Phys. Rev.* **B 41**, 12299 (1990).

[45] D. Mao and A. Kahn, M. Marsi and G. Margaritondo *Phys. Rev.* to be published.

[46] P. N. First, J.A. Stroscio, R. A. Dragoset, D. P. Pierce, and R. J Celotta *Phys. Rev. Lett.* **63**, 1416 (1989).

[47] W. J. Kaiser and L. D. Bell, *Phys. Rev. Lett.* **60**, 1406 (1988).

Crystal Structure

Crystal Structure Determination: The Synchrotron Radiation Advantage

Hans-Peter Weber

Institut de Cristallographie, BSP, Université de Lausanne,
CH-1015 Lausanne-Dorigny, Switzerland

SUMMARY:

The advent of synchrotron radiation (SR) has been of great benefit to solid state research.

For crystallography in particular, the high brightness of SR has permitted the use of single crystals of a smaller size (by one order of magnitude), and/or the study of biological samples with short beam lifetime. The ready availability of radiation in the high energy spectral range has reduced systematic errors in the data (like absorption and extinction) -thus greatly improving the data's quality. The small vertical divergence of SR has narrowed the width of reflected intensity profiles (by one order of magnitude), thereby permitting the detection of very weak reflections and the resolution of many more individual lines at high angle in powder spectra. The easy tunability of SR has allowed an almost deliberate choice of the magnitude of scattering factors, leading to an enhanced detectability of elements of interest.

Case histories serve to highlight these points.

INTRODUCTION

The following account endeavours to illustrate the remarkable advantages of synchrotron radiation (SR) over radiation from a conventional anode generator in the solution of crystallographical problems. It is an elementary review of a few such problems, the successful solution of which is due solely to the availability of SR. Several SR related crystallographic research topics have been left out. These include the application of SR to macromolecular crystallography and small-angle scattering. The latter was omitted as the emphasis in this chapter is on the determination of structure at atomic resolution; the former topic has recently been well reviewed elsewhere [1].

Still, it should be mentioned in passing that macromolecule crystallographers have been the first to take advantage of all the extraordinary features of SR (high intensity, easy tunability, low vertical divergence, time structure) to collect data and to determine the structure of, say, an enzyme or a DNA-drug complex. It is safe to say that macromolecular crystallography is probably the most hi-tech area of crystallography, as witnessed by the sophisticated techniques applied. E.g., genetic engineering is used to produce pure forms of the macromolecules in large lots, crystals are grown in low gravity to improve their quality, fast x-ray detectors have been developped expressly to speed up data collection for crystals with a short lifetime in an x-ray beam, and, finally, supercomputers have been applied extensively to the solution of their crystal structures because of the sheer size of the data sets. Of late, the most spectacular experimental achievement in this field has been the recording of an entire diffraction pattern of lysozyme with eight SR flashes of 120 picoseconds duration each [2]. In short, the space available in this chapter would not have done justice to this exciting field, even if restricted to just a few purely crystallographic questions.

Finally, let us point out that the following case histories are not arranged in order of importance, but only so as to give coherence to the survey. The particular contribution of SR to the solution of each illustrative example is summed up at the end of each section; to set this recapitulation apart from the rest

of the text it is printed in italics and emblazoned with a $\boxed{\text{SR}}$ symbol.

SCATTERING OF PHOTONS BY ATOMS, MOLECULES AND CRYSTALS

First, let us consider the scattering of a single photon by a lone electron. What do we measure? A typical scattering experiment always runs the following way. A stream of particles (photons, neutrons,etc) of flux I_0, generated by a source with a large spectral energy range, impinges on a target whose structure is of interest and is scattered by it (fig.1).

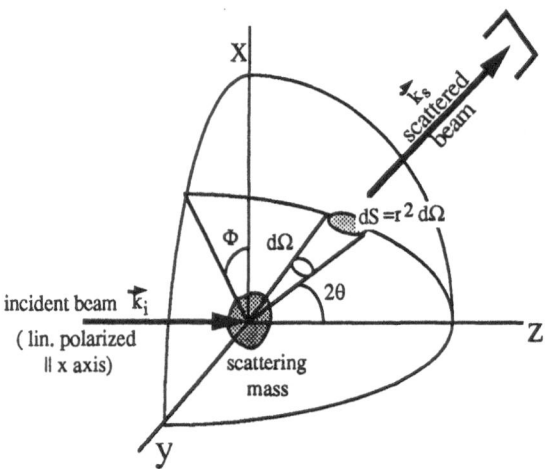

Fig. 1. Scattering geometry

All the information about the structure of the scattering mass (on any scale between semi-microscopic and atomic) lies in the differential scattering cross section, a very general quantity which is defined as

$$\frac{d\sigma}{d\Omega} = \frac{\text{\# of particles scattered per sec into solid angle } \Omega \text{ and direction } (\Theta,\phi)}{(\text{\# of photons per unit area and sec}) \; d\Omega} \qquad (1)$$

If one evaluates such cross-sections for x-radiation and scattering masses of increasing structural complexity, one arrives at the following results, assuming elastic scattering ($|\mathbf{k}_i| = |\mathbf{k}_s|$).

For a single free electron: $d\sigma/d\Omega = r_0{}^2 \, (\varepsilon_i \, \varepsilon_s)^2$ (Thomson scattering) (2)

where $r_o = e^2/mc^2 = 2.828 \; 10^{-5}\text{Å}$ and ε_i, ε_s are the polarization directions of the incident and scattered beam.

The scattering amplitude for an atom can be considered -to a first approximation- as the sum of the scattering amplitudes of all electrons shielding the atomic nucleus.

For an atom: $d\sigma/d\Omega = r_0{}^2 \, (\varepsilon_i \, \varepsilon_s)^2 \, |f(\mathbf{q})|^2$ (3)

where $f(\mathbf{q})$, the form factor of diffraction theory, describes the dependence of $d\sigma/d\Omega$ on the momentum transfer (or scattering vector) $\mathbf{q} = \mathbf{k}_s - \mathbf{k}_i$. The phase lag, $e^{i\mathbf{q}}$, for each electron is due the spatial distribution of the electrons in the atom. At large momentum transfer (i.e. for a large scattering angle), the form factor decreases to zero, as the waves scattered by the electrons interfere more and more destructively. The form factor is effectively a phase factor. Note in passing that the Fourier transform of the form factor is the atomic electron density.

Just as we added electrons to build an atom, taking into account the phase differences between each scattering center, we can add atoms to form a crystal. In contrast to the electrons in an atom, the atoms in a crystal are ordered in a regular fashion, and, as a result, incident particles are scattered only into particular directions. If we have a crystal, consisting of N unit cells,

For a crystal: $d\sigma/d\Omega = r_0{}^2 \, (2\pi)^3 \, (V/V_c{}^2) \sum \delta(\mathbf{q}\text{-}\mathbf{s}) \, |F(\mathbf{s})|^2 \, (\varepsilon_i \, \varepsilon_s)^2,$ (4)

where $F(\mathbf{s}) = \sum f_j(\mathbf{s}) \, e^{\mathbf{s} \, \mathbf{r}_j}$, with the summation over all atoms in the unit cell, and is called the structure factor.

It is evident from inspection of (4) that one measures the modulus of the structure factor, but not its phase. And so, in order to construct an electron density map of a structure, one has resort to some other techniques to model the phase. This is the phase problem of crystallography.

In a measurement, one samples a region $dS = r^2 d\Omega$ by scanning it with a detector of sufficiently large opening (see Fig.1). If the incident beam contains particles of only one energy, the observed integrated intensity relates to the total cross-section in the following manner

$$P(q) = \int i_0 (\lambda_0) \, \Delta\lambda_0 \, \sigma_{total}(q) \text{ abs ext } d\theta = \qquad (5)$$
$$\frac{I_0 \quad V \quad \lambda^3_0 \quad |F(q)|^2 \quad (\varepsilon_i \, \varepsilon_s)^2 \text{ abs} \qquad \qquad \text{ext}}{v^2_c \qquad \qquad \qquad \qquad 2\pi \sin 2\theta_B}$$

If the incident radiation is polychromatic, then:

$$P(q) = \int i_0 (\lambda) \, \sigma_{total}(q) \text{ abs}(\lambda) \text{ ext}(\lambda) \, d\lambda = \qquad (6)$$
$$\frac{I_0 (\lambda_B) \, V \quad \lambda^4_0 \quad |F(q)|^2 \quad (\varepsilon_i \, \varepsilon_s)^2 \quad \text{abs } (\lambda) \text{ ext}(\lambda)}{v^2_c \qquad \qquad \qquad \sin^2\theta_B}$$

These expressions - valid for any x-ray source- reveal well the parameters pertinent to the use of SR in crystallography.

They, first, clearly show that $P(q)$, besides being directly proportional to the incident intensity I_0, is also proportional to the volume of the crystal. Customarily, crystal sizes have ranged from 400 μm down 80 μm diameter, with a few measurements carried out on even smaller samples, when heavy (electron-rich) elements made up the bulk of the sample [3] [4]. SR is brighter by at least 4 orders of magnitude than radiation from the most powerful rotating anode x-ray generator; consequently, its use permits diffraction from crystals with a volume smaller by at least these orders of magnitude. Note that this tiny volume (~10 μm^3) is already close to the grain size of powders.

Secondly, $P(q)$ is shown to be inversely proportional to the square of the unit cell volume. In other words, for the same level of resolution, the incident intensity, I_0, is distributed in the case of a small mineral structure over, say, 1000 symmetry inequivalent reflections whilst in the case of a macromolecular structure like a protein this same incident intensity is scattered over >100'000 reflections. It is therefore not surprising that the use of SR with its phenomenal intensity improves the S/N of any reflection. This is also why macromolecular crystallographers have early on been at the forefront in the use of this type of radiation.

Thirdly, in the past, the λ^3 proportionality has been exploited to increase the available intensity at the expense of atomic resolution. This is why one used Cu radiation to collect data on structures with large unit cells. The use of SR obviates this stratagem.

Switching from mono- to polychromatic (or white) radiation boosts the diffracted intensity even further, as the whole spectrum is being exploited, with λ set to power of 4 instead of 3. This provides sufficient intensity for structural studies in time slices. And, just as important, $I_0(\lambda)$ can be calculated, in contrast to the spectrum of conventional sources. This last feature alone opens the way for the Laue method (fixed crystal-fixed detector) of collecting data.

To sum up, expressions (5) and (6) illustrate well the import of all the salient characteristics of SR to diffraction measurements - except for its small divergence. This last attribute benefits powder diffraction most, as we shall see later.

Akin to synchrotron radiation, the neutron spectrum of a steady-state reactor extends also continuously over a large energy spectrum. Except for a few trivial constants, the expressions (5) and (6) derived above for x-radiation apply also to neutrons. It seems appropriate, at this point, to emphasize that x-ray and neutron diffraction are complimentary (and not competing) tools. Whether one uses photons or neutrons to solve a structural problem is dictated more often than not by the nature of the problem at hand. In all applications where radiation intensity is the determining factor, SR will prevail as the radiation of choice, mainly because the scattering cross-section of nuclei for neutrons is so much weaker than the cross-section of electron shells for photons. Neutrons, on the other hand, offer several unique characteristics :

- They possess a magnetic moment which interacts with the electron spins. Although x-ray photons also interact with the electron spin, the significance of the interaction effects is orders of magnitude lower.

- Their energies are of the same order of magnitude as the energies of many atomic and molecular motions in solids. As such, neutrons are very suitable to probe collective excitations.

- The cross-section of hydrogen for neutrons is comparably much larger than for photons, and so neutrons remain indispensable for the localisation of hydrogen atoms.

RESONANT SCATTERING

The atomic form factor has been derived under the assumption that the electrons are free. The electrons, however, are bound tightly in different shells; they occupy either low-energy states or are excited to vacant states of higher energy by the absorption of a photon [*]. For particular frequencies, resonance will occur between the frequency of a bound electron and the frequency of the incident radiation, and absorption will occur; in an energy scan, this will reveal itself as a sharp increase of the absorption (absorption "edge"). These effects are

---

[*] Some of these vacant states are the continuum states of photoelectrons, the energies of which are modulated by the surrounding atoms (EXAFS).

small for light elements (below oxygen) and short wavelengths, and this is why they are often not accounted for in routine analyses of x-ray data. However, it is useful to remember that resonant scattering is always present in any x-ray experiment; it is just that away from the absorption edges its contribution is rather small, and was thus overlooked in the early days of x-ray diffraction.

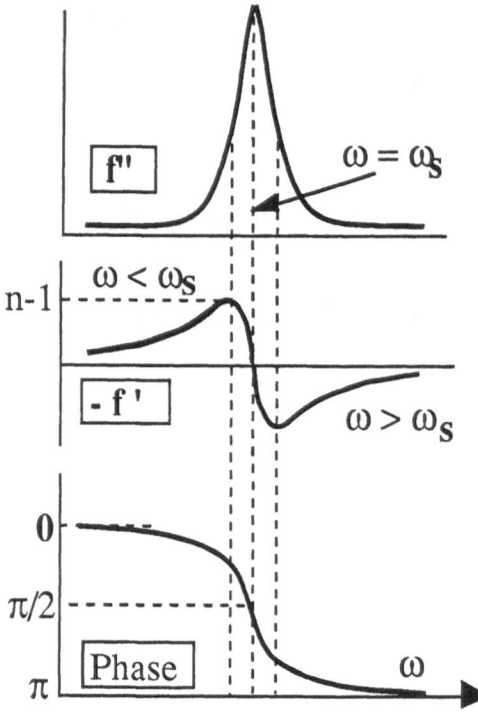

If we stay within the classical theory of dispersion, one can think of the atoms as consisting of electronic dipole oscillators, whose actual frequencies are those of the absorption edges of these K, L,... shells.

As the frequency of the incident x-ray photon approaches the energies of an absorption edge, the form factor changes both in magnitude and phase. To describe both these changes at once, the form factor needs to be expressed as a complex quantity,

$$f(q,\omega)=f_0(q)+f'(q,\omega)+i\ f''(q,\omega)\quad (7),$$

whose terms are shown in fig. 3 as a function of frequency.

f" is a complex number; the physical phenomena it describes are, however, real and in no way imaginary. The complex expression is just used to describe the magnitude and phase. If we draw the dipole scattering factor vs frequency, it is evident that we can distinguish

Fig. 2. Form factor.
The lower part of the figure depicts the frequency dependence of the phase

between three cases (see fig. 2). For $\omega < \omega_s$, the response of the oscillator is in phase with the driving wave. For $\omega > \omega_s$, the oscillator strongly opposes the driving force; this results in a phase shift of π. And at resonance, $\omega = \omega_s$, the maximum absorption is associated with an advancement of the phase by $\pi/2$ (together with a purely imaginary amplitude). Of essential import to SR crystallography is the frequency (or wavelength) dependence of the magnitude of f. It is used -as we shall see below- to enhance the scattering power of selected elements in a crystal structure.

Resonant scattering is often called anomalous scattering. The name is rather ill-chosen, and a relict of past experimental history. Anomalous scattering is namely the rule, and not the exception; this is the paradox. Calling anomalous scattering by its proper name, resonant scattering, allows one to see parallels to similar processes taking place with other forms of radiation like gamma-rays (also an e-m radiation) or with neutrons. When gamma rays are used , the resonance is known to all nuclear gamma-ray resonance, or popularly the Moessbauer effect. The nuclei undergoing such resonances are even more rare than for x-rays or neutrons. ^{57}Fe and ^{119}Sn (and some rare earths); these are not exactly elements common to most compounds. With neutrons, the nuclei subject to strong resonance are less numerous, unless one goes to very short wavelength (0.1Å).

POWDER DIFFRACTION AND ZEOLITES

Amongst mineral catalysts of technical importance, zeolites are particularly attractive materials to investigate; their synthesis is easily reproducible and the catalytically active surface can be well determined at the atomic level. Most of the research on zeolites of the past decade has been focussed on the synthesis of zeolites with novel structures and on the assessment of their technical usefulness. For those zeolites with promising catalytic properties, an inordinate amount of effort has been spent on optimizing these properties, often without a clear understanding of the catalytic reactions involved at the microscopic level.

The similarity between zeolites and enzyme proteins (one is in-, the other just organic) is not as far fetched as it may seem at first reading. Protein exhibit folds and zeolite contain cavities; both of these features select the reactant species. And, in both cases, the environment of the active sites is involved in complementary ways in the catalytic process. But there the similarity ends. Protein enzymes, on the one hand, consist of hundreds of atoms; their molecular weight ranges from 20'000 to 500'000 Daltons (unit cell parameters varying between 60 and 200 Å). Zeolites, on the other hand, are considerably smaller entities, and so it comes as a surprise that the enzymatic processes in the proteins are much better studied. There are two reasons for this seeming paradoxon. First, large (> 100 μm) single crystals of proteins have been easier to synthesize, despite the size of their molecular content. Secondly, due to the high symmetry of the zeolitic cavities, guest molecules -the reactants- are almost invariably oriented one way in one unit cell, another way in a neighbouring one. The resulting structural model is then the superposition of the content from all unit cells, and this makes it hard to model the positional probabilities of the guests.

Because of the difficulty of synthesizing zeolite single crystals of sufficient size and crystal quality, a great part of the structural studies on these materials have been done on powders. The difficulties one has encountered have been two-fold. First, the automatic indexing of the diffraction pattern from a compound with unknown unit cell crucially depends on the accuracy of the determination of the low-angle peak positions and their detection. And, secondly, the subsequent solution of the crystal structure hinges on the availability of a sufficiently large number of resolved unique reflections. The advent of SR, with its narrow collimation, has permitted to establish peak positions to better than 0.005° and to reduce the overlap between adjacent reflections, particularly at high angle. Due to this narrow collimation, the line width of reflections is not determined any more by the instrumental resolution of the diffractometer, but by the line broadening from the sample.

At the experimental level, the choice of scattering geometry and beam line configuration used to collect diffraction data vary greatly, depending on whether a particular experiment demands high intensity or high resolution . In the following, two of many possible configurations are presented (Fig 3). The

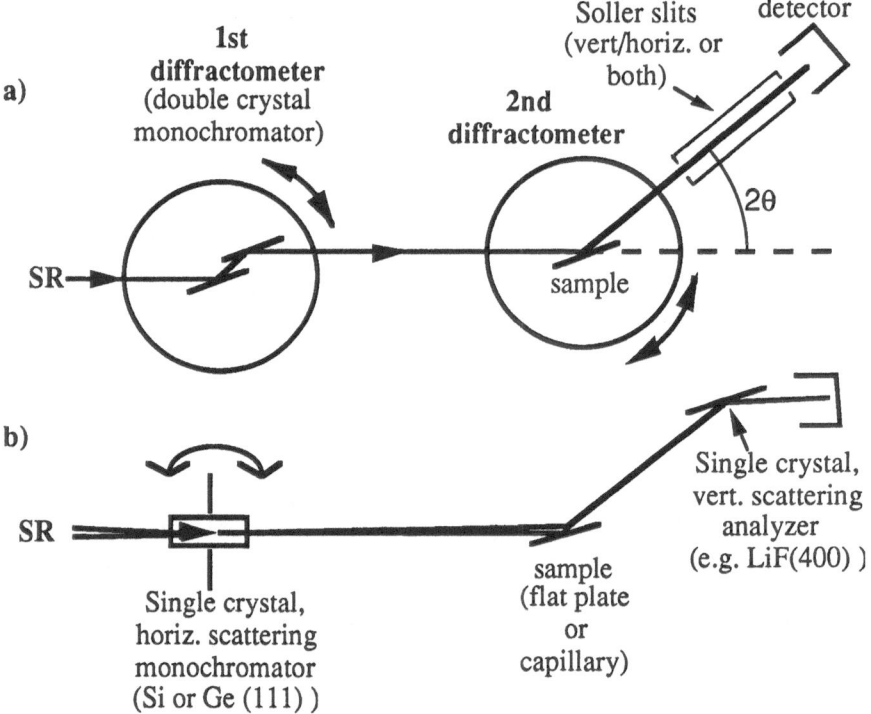

Fig. 3. Schematic layout of a high resolution powder diffractometer. Beam-defining slits have been omitted for the sake of clarity. **a)** Double-axis, parallel-beam geometry of Parrish-Hart [6] **b)** Triple-axis geometry of Cox et al.[5]

geometry is of the parallel-beam type due to the highly collimated nature of SR. As a result all diffracted x-rays are recorded simultaneously.

Also, in parallel-beam geometry sample angles and detector angles need not be coupled. The geometry may range from Θ: 2Θ to grazing incidence.

Right away, one obvious advantage of the high intensity and low divergence of SR is that one can afford to use perfect crystals as monochromators, with their concomitant higher energy resolution ($<10^{-3}$), but narrower angular acceptance for radiation. The pyrolytic graphite monochromators used in the laboratory are often perceived as yielding monochromatic radiation, but their energy resolution is at best several percent. And unfortunately, the broad band pass of conventional monochromators reflects not only the desired characteristic radiation, but also some Bremsstrahlung.

The configuration sketched in Fig. 3b, using an analyzer crystal, has the advantage of a very high accuracy in the determination of line positions (absence of any specimen displacement error) and of a superior peak-to-background ratio, due to the filtering-out of any fluorescent radiation. The perfect, vertically scattering analyser crystal functions as a very narrow receiving slit with a bandpass of a few eV. Instead of using an analyser crystal, one can use Soller slits to examine the diffracted beam; this yields higher intensities, but also a higher background. For applications requiring even higher incident intensities (high pressure cell, time-slicing), white radiation is used in conjunction with an energy-analyzing detector. The ab-initio structural determination of a zeolite from powder data is as good an example as one can get to illustrate the power of present-day structure-solving techniques on powder data, using SR. The zeolite in question, Sigma-2, is a new clathrasil[*] phase ($[Si_{64}O_{128}]$ $4C_{10}H_{17}N$), which exhibits enough structural complexity (large unit cell, $10.23 \times 10.23 \times 34.38$ Å^3, and disordered guest molecules) to serve as a good benchmark for the capabilities of the technique [7].

The sample material, filled in a capillary, was exposed to SR (λ=1.5468Å) at the National Synchrotron Light Source beam X13 (triple-axis geometry of Fig. 3b) of Brookhaven National Laboratory. A point detector scanned the resulting diffraction pattern in little less than 24 hours. Indexing of the peak positions revealed the cell to be tetragonal; from the observed, systematic absences of certain reflection types one inferred space groups I4$_1$md, I4bar2d or I4$_1$/amd. This result concurred with the outcome of an independent symmetry evaluation on the same material, using electron diffraction. The next step consisted in decomposing the observed pattern, i.e. assigning the correct intensities to each properly indexed peak. This step is crucial in the proper resolution of overlapping peaks, and it is here that SR -with its narrow collimation- has helped the most. Direct methods of structure solution were then used to reveal the structure, and the Rietveld technique to refine the atomic positions obtained (see

---

[*] i.e. its cavities are not connected by channels, and therefore its sorbent properties are poor.

fig.5). As straightforward as this may seem at first sight, one should not forget that a great deal of experience needed to be gained in each step involved, before one could bring the whole project to fruition.

Whole pattern refinement of the Rietveld type were first carried out on neutron data in the early 70's. A consensus soon established itself that such refinements were, in general, only good enough to refine atomic positions in already known structure types to a level of accuracy sufficient for some restricted purposes. High-angle peaks remained always unresolved, even if long wavelengths were used; background counts were thus difficult to estimate at these angles, and consequently atomic anisotropic displacement parameters, whose values depend on a clear definition of the intensity fall-off towards high angle, were of very marginal precision. With SR, however, the more than 10-fold narrower line width permits to resolve many more peaks at high angle (Cf fig. 11). Rietveld refinements on SR powder data are expected to lead to results rivalling single crystal studies of average accuracy. This is shown, e.g., in the remarkable work of Cernik et al. [8] on olivine, where -for the first time- the isotropic displacements parameters obtained are almost identical to the ones derived from single crystal data.

*In summary, we can conclude that the **narrow collimation** of SR has been the key factor in the progress of the technique of structural determination from powder data. We can expect SR to soon supplant neutrons as the radiation of choice in diffraction experiments on materials in powder form. However, neutron powder diffraction will retain its usefulness, if proper attention is paid to its unique properties . E.g., the generally small absorption of matter for neutrons makes for easy cryostat and furnace construction.*

*The advantage of the **wide tunability** range of SR wrt powder diffraction is that one can always choose a wavelength just slightly larger than the longest wavelength absorption edge of any elements in the sample at hand. For the ensuing lack of fluorescence increases the S/N ratio considerably.*

HIGH PRESSURE

Studies of physical properties at high pressure are central to science. It is just that the other intensive thermodynamic variable, T, is much easier to handle experimentally. Just like the temperature, pressure affects reactivity and reaction rates via its effect on excited electronic states. In contrast to temperature, pressure enhances multibody effects. As far as structural changes are concerned, an increase in pressure is analogous to a lowering of the temperature, with one important difference: experiments at very high pressures probe also the repulsive part (a consequence of the Pauli exclusion principle) of the interatomic potential

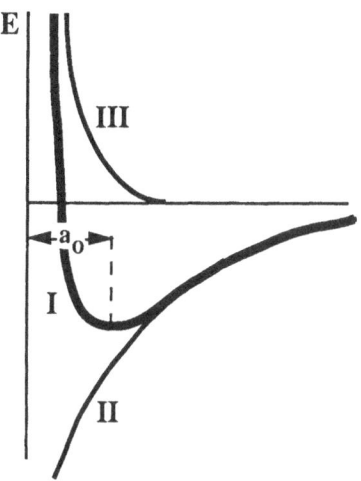

Fig. 4. Cohesive, total energy (I) consists of the electrostatic attraction between charged atoms (II) and the repulsive overlap between closed shells (III)

while a variation in temperature charts mainly the Coulombic, attractive part of this potential (see fig. 4). Successful molecular mechanics simulations of the (p,T) behaviour of materials require a knowledge of both terms, but only the Coulombic term is comparatively easily accessible, both theoretically and experimentally. A good analytical description of the repulsive term is harder to come by. Theoretically, several approaches exist (e.g., Modified Electron Gas model); empirically, only the modelling of the results from high pressure experiments yields realistic values for repulsive potentials. This is one of the many justifications for determining crystal structures at high pressure; more will arise in the following

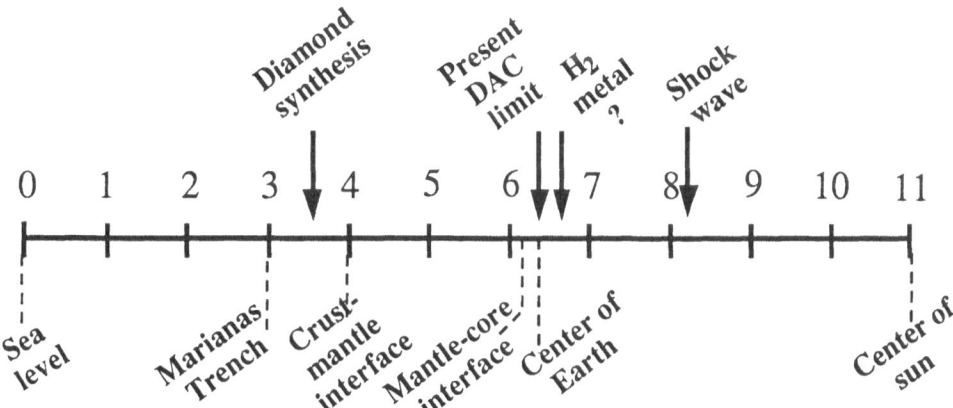

Fig. 5. Logarithmic pressure scale in bars (1 bar $=10^5$ N/m^2 (or pascal) = 0.9869 atm). Pressures attainable experimentally in the laboratory (shown above the pressure rule) are compared with the ones observed in nature (tabulated below).

paragraphs. Pressure is defined as force per unit area; pressure is thus either increased by augmenting the force or by reducing the area upon which this force acts. This simple fact has determined the evolution of pressure chambers, as we will see in the following. As a guide to one's intuition about pressure ranges, a pressure scale is shown in fig. 5. Development of high-pressure instrumentation have proceeded on two fronts. For measurements at high pressure on large bulk samples one resorts to large kiloton presses, up to 2 stories tall [9]. This has been the method of choice for the synthesis of quenchable high pressure phases and for the study of the macroscopic properties (conductivity, etc) of often multiphase assemblages. The limits of this technique are 100 kb and 1000 K.

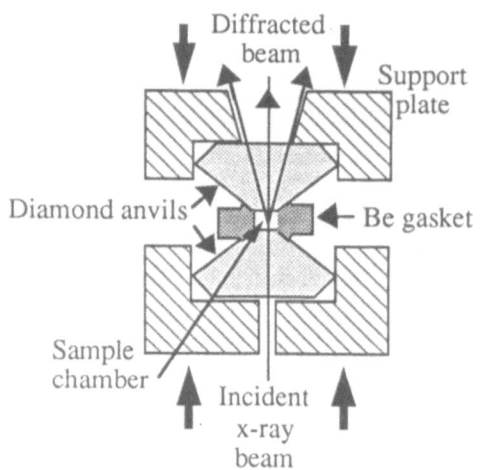

Fig 6. Diamond-anvil cell

For small (mm-size) samples, the high-pressure device of choice has been the diamond-anvil cell. The DAC is a mechanical apparatus wherein two opposed anvils made from diamond are pushed together, with the sample to be studied squeezed in between. There are several designs for such a cell, depending mostly on the observational technique used to monitor the reaction of the sample: spectroscopic ones like Raman, Moessbauer and NMR, or diffraction techniques. The principle is sketched in the figure 6.

DAC's also vary in design as a function of the pressure range for which they are built. Below 200 kbars (20 GPa) the alignment of the anvils is not too critical; any misalignment is absorbed by the gasket. At higher pressures, however, the anvils have to be aligned very carefully, and this precise alignment preserved up to maximum loading. At very high pressure, in the megabar range, the diamond anvils begin to deform elastically and the gasket plastically, i.e. irreversibly. With further pressure increase the diamonds fail, because of the large pressure gradient at the edge of the culet. The remedy to this problem is to bevel the culet's rim, to cut it at a slant. Though one gets rid of premature failures upon loading, the anvils still crack upon release from high-pressure because the diamond anvils regain their original shape while the gaskets do not. There is actually no theoretical limit to the maximum pressure attainable in a diamond cell. A phase transition is predicted above 14 Mb. However, this transition does not necessarily represent a limit; diamond may behave metastably, as it does in the graphite phase field. Previously, all pressure frontiers have been design-related. However, the present limit is more of a practical nature.

Crystallographic experiments at high pressure have to date been mainly the preserve of physicists and experimentally inclined mineralogists. Gearing up for a high pressure experiment - even with a conceptually simple pressure device like a DAC - is not as routine as dialing the temperature on a cryostat or on a furnace. Measuring the generally low intensity of X-rays diffracted from a sample contained in a high-pressure device is very time consuming when carried out with radiation from a conventional X-ray generator. The main difficulties are the long absorption paths through diamond anvils and pressure confining metal gaskets as well as the limitations imposed to the crystal size (< 100 μ) due to the minute sample volume available.

One of the simplest diffraction experiments carried out with a DAC is the determination of cell constants as a function of pressure (and temperature), from which the equation-of-state (EOS) is derived. The increase in accuracy obtainable with SR (many more peaks at high angle due to higher intensity of SR beam; better resolution because of the low divergence of the SR beam) permits one to derive the EOS with such an accuracy that it can be used, by comparison, to test the validity of theoretical models for material behaviour.

Another exciting field of study is the investigation of elemental gases in their solid state. The properties of hydrogen, in particular, the most common element in the universe, are of fundamental importance in modelling the interior of giant planets or in charting the early evolution of the universe). A knowledge of its EOS in the presently reachable pressure regime permits one to predict its behaviour at ultrahigh pressures. At such pressures, , when the hydrogen molecules are expected to dissociate to their atomic constituents, solid hydrogen is foreseen to become the simplest conducting metal (maybe even a superconductor). Intuitively, one would expect, in diatomics, the inter- and intramolecular bonds to become of comparable strength at high pressures, with both types of bonds becoming chemically undistinguishable. The first expectation turns out to be true; the second one is not realized. Bonds retain their individuality at high pressure, and molecules their identity.

The structure solid H_2 adopts over the entire phase space is also of interest to theoreticians because solid H_2, together with solid He^4 and He^3, are the prototype quantum solids. These are solids where the amplitude of the zero-point lattice vibration amounts to a considerable fraction of the cell constant, due to the small molecular masses and the weak interaction between molecules. The structure of solid hydrogen at ambient pressure is known from low temperature experiments, it is hcp with the space group $P6_3/mmc$ [10]. This early structure determination was no mean feat, in view of the fact that x-rays are scattered by electrons, and that each hydrogen atom has only one electron. Collecting x-ray intensities on solid H_2 at high pressure is even more of a challenge because the weak scattering

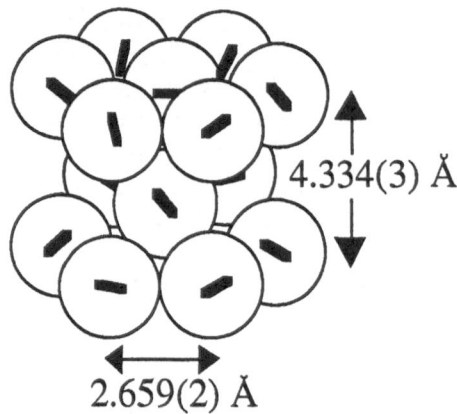

4.334(3) Å

2.659(2) Å

Fig. 7. The crystal structure of solid H_2 at 54 kbars. The H_2 dumbbells are represented by bars.

from the tiny hydrogen single crystal is drowned out by the scattering from its high-Z (DAC) environment.

In a first experiment at 54 kbars, on a diffractometer with a conventional x-ray tube, it took the dauntless experimentalists [11] about 4 days to find the first weak refl. (50 cps as compared to a background count of 15 cps). And this was actually the strongest reflection predicted for the HCP structure! A total for 8 reflections (and their symmetry equivalents) were measured. From Raman spectra on solid H_2 at the same pressure it had already been deduced (only transitions between

purely rotational states were observed) that all H_2 molecules are rotating and must thus be centered on the nodes of the Bravais lattice (see fig. 7). Thus, the intermolecular distances were given by the unit cell dimensions. The axial ratio oberved at 54 kbar was 1.630(3), close to the ideal value of 1.633. In the structural refinement, the H_2 dumbbells were assumed to be freely rotating; RMS thermal displacement amounted to 0.25(3) Å as compared to an intermolecular distance of 2.659(2) Å. Because higher pressures are only attainable with a reduction in the sample volume and a concomitant loss in diffraction intensity, experimentalists moved to a SR facility. They were then able to determine the crystal structure of solid H_2 at four additional pressures up to 265 kbars [12]. The structure type remains the same; the major difference lies in an unanticipated decrease of the axial ratio to 1.594(5), indicating an increase in the anisotropy of the solid. A complete orientational ordering of the H_2 molecules would have yielded a new phase with Pa3 cubic structure; so it is more likely that what is observed was an incipient preferred orientation of the rotation axis parallel to the c-axis.

The range of compounds investigated as a function of pressure in the last 10-15 years extends from susbtances with simple structures like the one just described to the most complex macromolecules. The effect of pressure on, say, proteins is just as spectacular - and relevant to their biological functions - as it is in the mineral world. We know that the application of pressure deactivates enzymes, viruses and toxins, and reverses narcosis induced by alcohol. We also know that sharks easily adapt to life at great depths (to 1 kbar water pressure). The elucidation of these effects at the molecular level, however, is not trivial. The main problem is that proteins crystallize in a peculiar state of condensed matter:

the crystals consist usually of 40-60 percent of well-ordered amino-acids with the balance in weight made up of disordered solvent molecules like water and dissolved salts. There are indeed very few direct contacts between protein molecules. Much of the solvent has kept the same properties as in bulk form; its state is liquid, not solid. Thus, when one compresses a protein and measures its cell dimensions, the result one obtains is not indicative of the real compressibility of the protein molecule. One has to determine its full 3D structure. And this has not yet been done, although hen egg lysozyme has been squeezed to 2 kbars. It will be interesting to see how the conformation of the subunits of a protein vary with pressure. A protein may possess many conformational substates, and pressure is expected to alter both the relative population of the substates and the functional properties of each substate.

Another exciting application of the DAC, still in its infancy, is the exploration of structural changes in incommensurate crystals as a function of p; sofar their behaviour has been investigated only as a function of T. The introduction of the second intensive variable, p, has recently revealed an amazing structural complexity in a compound previously thought as simple [13]. It can generally be said that models describing the behaviour of incommensurate crystals can only conclusively be tested if a complete (p,T) phase diagram has been structurally mapped. Sofar, compounds of interest have included mostly metal-organics, and it is safe to assume that SR with its high intensity will contribute decisively to a better understanding of the role of their light-atom moieties.

In summary, one can say that the use of SR for the structural determination of solids at high pressure is essential for the experiment to succeed because :

*- Higher pressures can only be obtained with smaller sample volumes. The ensuing reduction in scattered intensity is more than compensated by the **higher intensity** of SR.*

*- The already **narrow collimation** of SR can be reduced even further, down to 5 µm, to illuminate only a small volume within the specimen where the pressure gradient is small. The same narrow collimation also avoids striking components of the DAC, and this increases the S/N ratio.*

*- The availability of very short wavelengths in the **SR spectrum**, with their higher penetration depths, lessens the severity of the absorption correction.*

*- Structural studies at high pressure is one field of SR crystallography where **white radiation** is being used to great advantage in conjunction with energy-dispersive diffraction techniques; only this combination permits the foray into the megabar range.*

HIGH T_C SUPERCONDUCTOR

Structurally speaking, all high T_C cuprates crystallize in structures related to the perovskite structure, albeit in some cases only very remotely so. Thirteen structure types have sofar been recognized, three are listed below, with their main structural characteristics briefly summed up in the following paragraph.

Composition	Structure type	T_c
La_{2-x} $(Ca,Sr,Ba)_x$ Cu O_4,	$\{K_2\,Ni\,F_4\}$ type or "211"	25-40K
$Ln\ Ba_2\ Cu_3\ O_{7-\delta}$ (where Ln = Y, Nd,Sm,Eu ,..Yb; $\delta \sim 1.3$)	"YBCO" or "123"	~ 90K
$A_2\ Ca_{n+1-x}\ B_x\ Cu_n\ O_{2+4}$, (where A= Bi or Tl; B=Sr or Ba; n=1÷3)	"2212"	112K

The perovskite structure, with the general formula ABO_3, is essentially a framework of BO_6 octahedra linked at their corners; the cation A, always large in size, occupies the central cavity (fig. 8). The structure of "123", one of the most thoroughly examined oxides, is essentially a perovskite triple-decker (e.g., A=BaYBa ; B = Cu3) with the cavity cations ordered in the sequence -Ba-Y-Ba-Ba-Y-Ba-Ba- . However, $YBaCuO_{7-\delta}$ differs from this ideal structure (with the structural formula $A_3B_3O_9$) by the absence of up to three oxygen atoms (see fig. 8).

The concentration of oxygen vacancies is coupled to the charge on the Cu atoms. A raise in T_c seems to always go hand in hand with an increase of the structural complexity. A good case in point is the structure of "2212"with the formula A_2 $(Ca_{n+1-x}\ B_x)\ Cu_n\ O_{2+4}$,which displays an even lesser resemblance to perovskite (see fig. 9). Essentially, this structure consists of sheets of corner-sharing, square-planar CuO_4. Two of these sheets always stack to a double sheet, the apices of the CuO_5 pyramids facing away from each other. Intercalated between these sheets one finds mainly Ca atoms, with a few Sr atoms substituting for Ca. The majority of the Sr atoms, however, are embedded between the apices of the CuO_5 pyramids. And, finally, between these double sheets of stoichiometry CuO_2, one finds double layers of bismuth cations.

Common to all these compounds is their quasi two-dimensional, layer-like structures (with varying stacking sequence), the omnipresence of covalent Cu-O bonds and the fact that the cationic species, their concentration and/or the oxygen

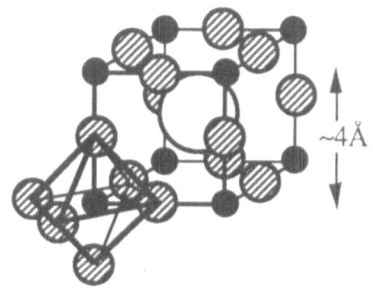

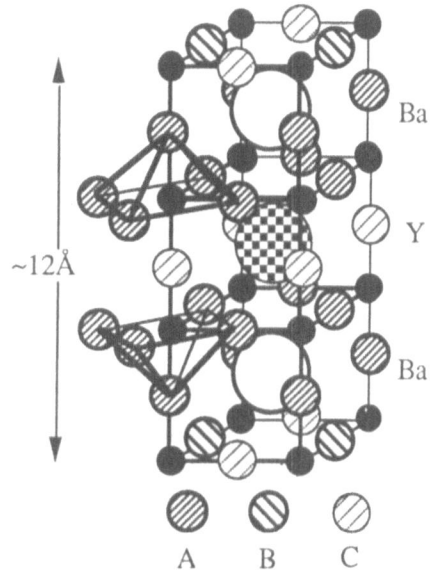

Fig.8 (top) Perovskite structure, ABO3. A (unfilled) atoms: Y, Tl, Ba, K,...; **B** (black) atoms: Bi, Cu, Ti,...; **O** (hatched) atoms: oxygen.
(right) Structure of YBaCuO7-δ ("123"), represented as stacking of 3 perovskite-type modules. Type A oxygen atoms are always present. Type C oxygen atoms (on the vertical edge of the yttrium cube) are always absent, leading to the stoichiometry YBaCuO7. If type B oxygen atoms are also lacking, the stoichiometry is YBaCuO6; this oxygen poorer variant is not as good a superconductor.

stoichiometry determine 1) the ratio of the number of copper cations with formal charges +2 to the ones with +3 and 2) the symmetry of their structures (whether in the ideal or distorted form). This charge ratio seems to control the onset of superconductivity.

The structural determination of these complex, mostly non-stoichiometric compounds have been hampered by the facts that:

-They are difficult to grow as large single crystals and, thus, one has to resort to powder diffraction for structure determination, with its concomitant loss in precision.

-In the superconducting state the crystals are twinned.

-Electron-rich (rare earths, Cu) and electron-poor elements (like oxygen) occur together. X-ray diffraction will sense mostly the high scattering power from the former. This would not be so troublesome if the oxygen atoms were not thought to be key players in the mechanism of superconduction. An occasional solution to this problem has been the use of neutron diffraction where the

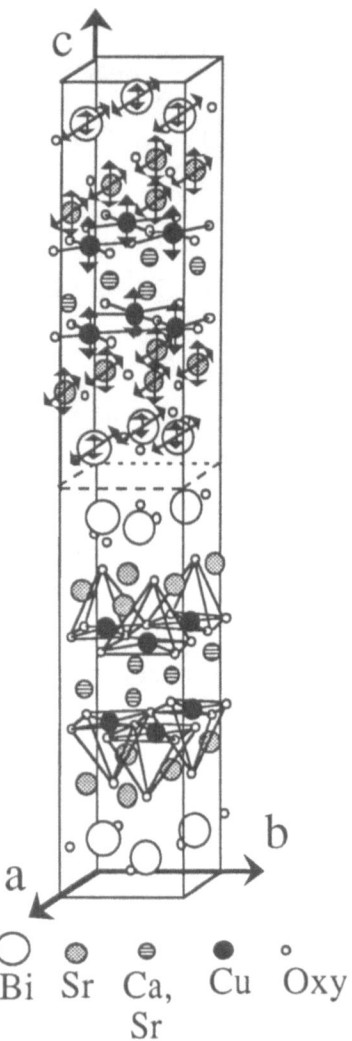

scattering power is not linearly dependent on the atomic number. But the low flux at present-day steady-state reactors require the use of large (several mm^3) samples. The use of SR fortunately remedies this problem. Because SR offers an intense continuum of wavelengths, one is able to accentuate -by tuning in to absorption edges- the scattering power of any element (heavier than Ca) which happens to be a relatively weak contributor to the total scattering in a structure, but is of great structural interest. This is illustrated by the following example.

In the system Bi/Sr/Ca/Cu/O the existence of superconducting phases was first established in explorations of the phase diagram of that system [14]. Many of the samples synthesized exhibited transition temperatures between 85 and 95K. Eventually, a new superconducting phase with the composition $Bi_2Sr_{3-x}Ca_xCu_2O_{8+y}$, the structure type "2212" already described above and an onset of T_c at 116K was isolated from the sample mix. Spurned on by the high T_c, materials scientists investigated its structure in great details, still hoping to find the elusive structural clue to superconductivity. Examination of the first platy crystals with the electron microscope revealed the presence of

Fig. 9. Structure of "2212" or Bi₂Sr₂CaCu₂O₈. In the upper half of the unit cell, arrows show the main displacive modulation directions. The lower half displays the polyhedral packing.

superstructure reflections along the a^* axis. The existence of satellite reflections pointed to either a displacive (positional shifts) or a substitutional (chemical occupancy) modulation.

The satellite reflections were displaced by $q= +/- 0.21$ a^* from the fundamental diffraction spots. $1/q$ not being a multiple of one, the modulation is incommensurate with the fundamental Bravais lattice. This is where matters rested at first. Eventually, several labs succeeded in synthesizing larger single crystals of this composition, and the structure was solved and refined - at first, just the average structure. When the composition refined from the analysis of the average structure did not agree with the results from chemical analysis - and both methods had been expertly handled -, a group led by Coppens et al. [15] had a closer look at the structure. The questions which needed to be answered were:

- What is the exact stoichiometry of this compound ? A knowledge of the total composition mattered because one wanted to know whether the Sr content was governing the oxidation state of Cu - in contrast to YBCO, where the oxygen concentration determines the charge on Cu (In the compund at hand, all oxygen sites are fully occupied).

- How are the cations distributed over the several possible sites ? A knowledge of the exact cation distribution was important because a substitution of trivalent Bi for divalent Ca,Sr changes the hole concentration in the CuO_2 plane.

-   What is the import of the observed modulation to the onset of superconductivity?

An investigation with synchrotron radiation commended itself for the following reasons:

- Even though the cations (Bi, Ca, and Sr) differed widely in their scattering power, a detection of their shared presence on any site in concentration lower than 10% (i.e. at the level of precision required to establish the accurate stoichiometry) could not be hoped for with the use of a standard laboratory X-ray source. The site occupancies correlate too well with the atomic displacement parameters.

- The high order satellite reflections, first observed with electron microscopy and notoriously weak in intensity, are only well observed with SR because of the latter's high intensity and narrow collimation.

In our discussion of resonant scattering we have seen that the scattering power is wavelength-dependent, and varies particularly strongly at absorption edges. In this particular compound, Bi has an absorption edge which is easily reachable. If one tunes to it, Bi will contribute to the measured intensities out of proportion to the other elements (unless their absorption edges happen to lie close by). f' varies from -21.15 at the absorption edge (0.924 Å) to -9.51 at 0.960 Å, while f" varies little in this wavelength range (3.96 and 4.21 resp. at the two wavelengths). Although the space group of "2212" is Amaa, the structure is almost centrosymmetric, and the imaginary part of the scattering factor is of little import to the experiment. The scattering factors of the other elements did not vary much in the wavelength range of interest, a crucial point in the strategy for the subsequent analysis.

From data sets collected at the two wavelengths the investigators computed at first electron density maps. By subtracting maps calculated with data collected off Bi resonance from the one measured at resonance, they were able to single out the Bi distribution. Bi was found in 3 different crystallographic sites, not just the nominal Bi site -although the residual peaks were quite small. Keeping positional and displacement parameters at the values obtained previously for the average structure, they were able to refine these concentrations (two of them of the order of 5%) and to determine the exact stoichiometry.

When the new distribution from SR data was checked against the sealed tube data, the agreement neither improved nor worsened, indicating the latter data to be quite insensitive to such modelling. The new stoichiometry, $Bi_{1.91}Sr_{1.72}Ca_{0.80}Cu_2O_8$, leads to an average formal charge of 2.62(6) on the Cu atoms. The fact that the Ca + Sr content adds up to less than 3 is alone responsible for this and the creation of holes in the CuO_2 layers. No extra oxygen was needed.

The analysis of the reflections with non-commensurate indices [16] [17] showed the modulation to be mainly displacive instead of subtitutional, and a detailed analysis of the data revealed that all atomic species present in the structure were modulated. The Bi atoms reveal a large amplitude modulation along the a axis with a smaller one along c. For Sr, these components are about equal. The Cu atoms, on the other hand, show only a large modulation along c - which is understandable in view of the strong covalent bonding in the CuO perpendicular to this axis (See Fig. 7).

The easy tunability of SR, allowing a deliberate choice of scattering power for certain elements, and its high intensity (weak satellite reflections; small crystals) were the key factors in the success of the above investigation.

ELECTROSTATIC PROPERTIES FROM ACCURATE, HIGH RESOLUTION DATA

The electron density observed in a structure is derived via Fourier transform from intensities determined in a diffraction experiment. This density is thermally smeared out, being the convolution of the electron positional probability density of each atom with the nuclear probability density function of the same atom. With state-of-the-art x-ray data it is nowadays feasible to parametrize the first quantity, the electron density, and to reliably derive from it chemically meaningful electrostatic properties like the potential , the atomic charges (albeit in a model dependent form) and the charge redistribution taking place upon bonding [18].

Similarly, the nuclear probability density functions are determined with ever increasing accuracy and precision. They contain information on the motion of atoms in the crystal and on the forces governing these motions. An understanding of atomic motions is a prerequisite for an appreciation of chemical transformations.

In the past two decades, much effort has been spent on the careful determination and interpretation of each of these two probability terms [19] [20].

The use of SR to collect diffraction data at a higher level of accuracy commends itself for the following reasons:

- The natural collimation and the high brightness of SR results in a very significant increase of the signal-to-noise ratio of observed intensities (both Bragg and diffuse). The contribution from thermal diffuse scattering (TDS) to a reflection profile is easier to separate graphically, because the TDS peak is much wider than the Bragg peak. Reflection profiles are typically an order of magnitude narrower than those measured with radiation from a sealed tube, and, so, weak reflections rise out of the background. Pseudo-forbidden reflections, one kind of weak reflection, are often encountered in accurate electron density studies. The (222) reflection in the diamond type structure (Si, Ge, C,...) is a case in point. Their illegal status arises from the fact that the 232 space groups (and the extinction conditions deduced therefrom) have been derived on the assumption that atoms are perfect spheres. Often, however, the slight asphericities due to either chemical bonding and/or anharmonic motion do not follow the local symmetry of a particular atom's environment. More often than not, the real local symmetry this atom displays is 1; the systematic absences associated with the higher local symmetry are not present any more, and additional, mostly very weak reflections are henceforth observed. Another kind of weak reflection, already discussed in the section on high T_c superconductors, are the satellite reflections which are due to modulated departures (of the substitutional or displacive kinds) from a particular structure type. As these deviations are mostly quite small, the additional reflections they give rise to are weak, and only SR permits their observation up to high orders. And finally, Bragg reflections at high diffraction angle are always weak because of the strong fall-off of the form factor towards high angles and because for most scattering geometries one is far off focusing condition. Although weak, these reflections are essential to resolve the detailed features of the electron density.

- Systematic sources of error like extinction or absorption become less severe. Extinction[+) decreases as crystals become smaller and/or wavelengths shorter . At high energies the properties of X-rays are quite different from those at, say, 8 KeV (1.5Å). With photons at 30 KeV (0.4Å) absorption is drastically reduced (factor of >20). These experimental advantages were first demonstrated

---

[+) Extinction describes the degree to which a crystal approaches perfection in its reflection behavior. In a perfect crystal, where no irregularities like dislocations disturb the macroscopic fabric, an incident beam never reaches the inner part of the crystal because it has been already reflected out on its path through the crystal or because of multiple reflection. This specious absorption adds onto the regular photoelectric absorption along the Bragg reflection direction.

on CaF_2 , a compound reputed for its fierce extinction [4] [21]. In a project sponsored by the American Crystallographic Association, crystals of CaF_2 had been passed around different x-ray laboratories in order to compare the quality of each laboratory's experimental technique. In retrospect, the choice of substance turned out to be an unhappy one. Because of the -then and now- unsolved modelling for extinction, this round-robin exchange shed less light on the level of experimental skills of the participating labs than on their computational approach to extinction in the subsequent structural refinement. The recent SR experiment was carried out on a crystal with an average dimension of 6 μm (see fig. 10). The refinement of this simple structure showed the SR data to be quasi extinction-free. Also, even though data had been collected previously on small crystals of a similar size (~30 μm), the lack of heavy elements in CaF_2 was a far better test of the feasibility of collecting data on tiny crystals.

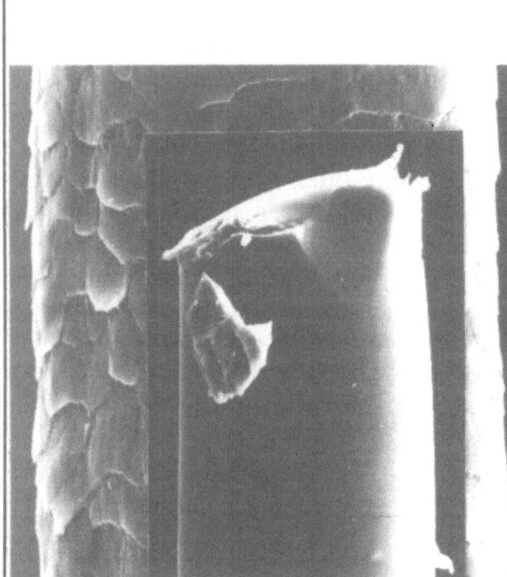

Fig. 10. CaF_2 single crystal with an average dimension of 6 μm, mounted on a glass fiber. A human hair is shown in the background, for scale purposes only.

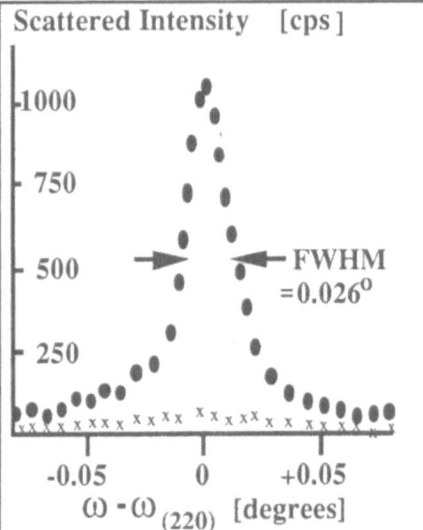

Fig. 11. ω -(or rocking) scan of the (220) reflection of the CaF_2 single crystal. Profile recorded with SR is represented with filled ellipsoids; x 's outline a scan performed with x-radiation from a sealed tube. FWHM for SR data is ~10 times smaller than for sealed tube data.

- The use of a short wavelength (and of a higher brightness) permits collection of (significant) data to much higher resolution resolution than previously feasible in a university laboratory. At the higher resolution, the contribution from the valence electrons peter out, and the main scatterers are the core electrons. While, in the past, accurate data at such high resolution has been next to impossible to come by, its recent availability at SR sources will lead to a better description of the electron density, in particular of the core electron density, which is yet to be shown to deviate significantly from sphericity.

Few accurate electron density studies have been performed with SR sofar [22] [23]. These have been mostly feasibility studies and have restricted themselves to proving that the quality of the SR data was at least as good as the one obtained earlier with x-rays from a sealed tube. No novel information of a chemical or physical nature has emerged from these studies and we need not go into details. Within the topics covered by this survey, this is probably the field which has taken least advantage from the availability of SR.

THE STRUCTURE OF SURFACES

Sofar, we have been concerned with the detailed atomic arrangement in bulk crystals, and we have assumed that the structure observed extends throughout the entire bulk. We neglected the fact that the bounding surfaces of these solids are expected to adopt a different structure than the bulk, because of the one-sided bonding of the surface atoms to the bulk of the crystal ("dangling bonds"). The surface layer tries to minimize its free energy by recrystallizing ("reconstructing") with a 2-D periodicity different from the periodicity expected for a simple termination of the bulk, and the top atom layer is often buckled (inward relaxation). This 2-dimensional structure will also vary as a function of temperature. Far below melting, surfaces adopt well-defined structures as manifested by clean facets, whereas they become increasingly rough as the temperature approaches melting.

Experimental methods like low-energy electron diffraction (LEED) and ion-scattering spectroscopy have confirmed this view, and until the mid-seventies these methods reigned unsurpassed by any x-ray technique. The reason was that hard x-rays (> 10 keV; < 1Å) are very penetrating (of the order of several μm in perfect and several mm in imperfect crystals). To probe only the topmost layers (= surface), it is necessary to select an x-ray method which detects only deviations from the bulk structure or, in the case of diffraction, to either 1) limit penetration to the top layers by striking the surface of the specimen at a grazing angle within the range (< 0.5 deg) for which total reflection occurs or 2) to restrict one's observations to surface-specific reflections (superlattice reflections from reconstructed surfaces). When x-rays impinge on the surface at an angle larger than for total reflection, the contribution of the surface layers to the total Bragg scattering is proportionally so small as to be insignificant.

The need to explain and control, at the atomic level, industrial applications like catalysis, crystal growth, sintering, wetting and adhesion has provided the

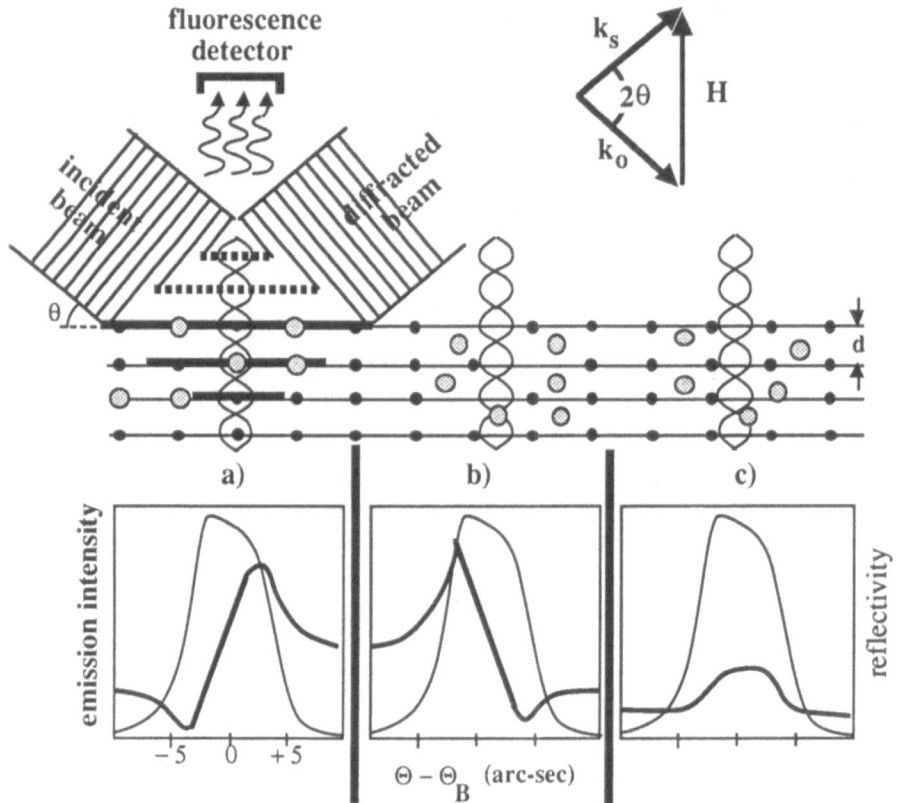

Fig. 12. Schematic illustration of for XSW fluorescence experiment.
The interference pattern arising from the superposition of plane waves with incident wave vector k_0 and k_s is shown to have a periodicity ($= \lambda/2 \sin\theta$) equivalent to the periodicity of the diffraction plane, due to Bragg's law for the bulk crystal yielding also $\lambda/2 \sin\theta_B$ or $k_0 + H = k_s$ (see inset top right). H is the reciprocal lattice vector for the diffraction plane with spacing $d = 1/|H|$. As the Bragg angle is scanned, the phase of the XSW relative to the regular diffraction planes and the plane of the impurity atoms (shaded in grey) manifest itself in different fluorescence peak shapes. We differentiate between 3 cases: a) Impurity atoms lie in regular diffraction planes, i.e. they substitute for regular atoms, b) Impurity atoms are located on well-defined interstitial sites, and c) Impurity atoms are distributed incoherently between diffraction planes.

The heavy lines (bottom panels) represent the fluorescence intensity, the lighter ones ,with the top-hat shape, the Darwin-Prins reflectivity (or "rocking curve").

impetus for the investigation of surface phenomena, and in particular the determination of the surface structure.

This section introduces two relatively novel x-ray techniques which also permit the study of surfaces, but surpass by far the above mentionned techniques, and are heavily dependent on the availability of SR : x-ray standing wave (XSW) fluorescence and grazing-incidence (GI) surface diffraction .

According to the dynamical theory of diffraction, a standing wave field is excited when -in the case of Bragg reflection- incident and diffracted, coherent travel-ling plane waves interfere in a nearly perfect crystal. The excited wave field vibrates in the direction perpendicular to the family of reflecting planes and takes on the same periodicity as these planes. As the crystal is rocked through the diffraction peak, the phase of the XSW changes by π wrt to the diffraction planes. On the low angle side of the rocking curve the nodes of the standing wave the planes of maximum electron density- , and on the high angle side it is the antinodes which do*). The photoelectric cross-section is proportional to the intensity of the E-field; fluorescence thus rises as the antinodes of the XSW move across the atoms of interest. Atoms located at the antinodes of the standing electric field emit fluorescence x-rays and secondary (photo and Auger) electrons (see fig. 12). By correlating the position of the antinodes with this secondary emission, it becomes possible to locate the excited atoms along the direction normal to the diffraction planes, with an accuracy of a few percents of the cell constant. In order to locate the atoms in three dimensions, one triangulates them by observing the fluorescence from other diffraction planes, not coplanar with the surface.

As conceived initially by Batterman [24], the XSW fluorescence technique was only surface-sensitive in the sense that the impurity atoms it probed were known to be located only in the topmost layer. The penetration of the XSW (1/extinction length) was eventually shown to depend on the collimation at the fluorescence detector, and is of the order of 1 μm (or several thousand layers) [25]. Later on, standing waves were also shown to arise not only below the surface, but also above, and so the XSW technique applies itself also to the location of atoms adsorbed to a surface [26].

The application of the technique have been manifold. Originally, it was used to decide whether impurity atoms implanted in semiconductors did substitute for regular atoms or preferred to reside on interstitial sites. Recently [27], its range of application has been extended to determine the structure of the interface between an epitaxially grown film and its substrate. In the process, the XSW fluorescence technique has proven itself to be one of the few techniques capable of probing a buried interface at high resolution.

---

*) Let us note in passing that standing waves are also at the root of other phenomena in dynamic diffraction. The anomalous transmission of x-rays in perfect crystals, the Borrmann effect, is such an example.

One such example is the study of the metal-semiconductor interface encountered in micro-electronic devices. A more precise knowledge of the structure of such interfaces is likely to shed some light on the formation of Shottky barriers as these ohmic contacts are finding increasing industrial use. Metal silicides ($NiSi_2$, $CoSi_2$,...) crystallize in the cubic CaF_2 structure, and are one case, type A, all crystal axes axes align with the substrate; for type B, the under which one type grows to the exclusion of the other. The bulk mismatch is known to grow epitaxially on Si (111) surfaces with two different orientations. In epilayer is rotated 180 degrees around [111]. One is able to control the conditions of the order of 1.2 % for $CoSi_2$ on Si (111). The central question was the interfacial structure. From simple model building (see fig. 13) it was obvious that there were two possibilities. To decide between the two, XSW were generated by Bragg reflection off the Si (111), and the positions of the Co atoms relative to the nodal planes above the Si surface determined from fluorescence measurements [28]. The example shown in fig.13 a) turned out to be the correct one. In the case of $NiSi_2$ the interfacial structure ist just the opposite (fig. 13 b). Ni is then seven-fold coordinated.

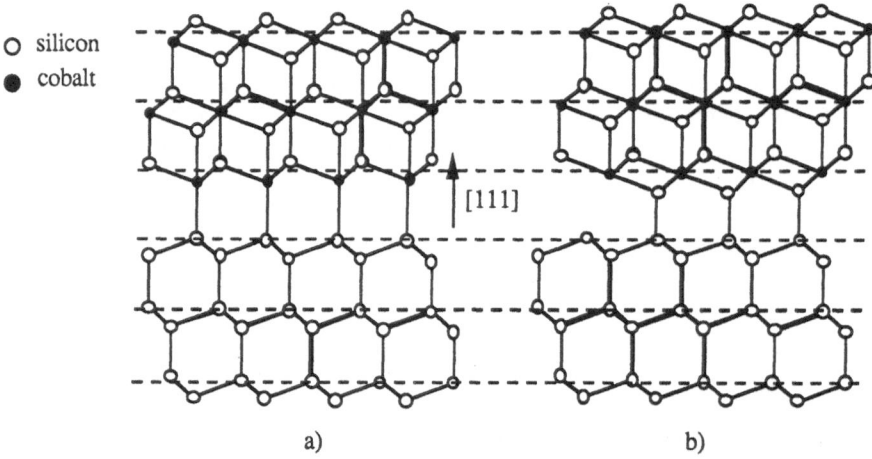

Fig. 13. Models of the $CoSi_2$ on Si (111) interface. The orientation of the epilayer wrt to the substrate is the same in both cases and of type A. Case a: Dangling bonds of Si substrate link to Co atoms in the 1st epilayer. Case b: These same Si atoms bond instead to Si atoms of the epilayer. Dashed lines serve as guidelines for the eye.

Another example of the power of XSW fluorescence measurements is the determination of the thermal motion of a Br submonolayer adsorbed on a Ge (111) surface [29]. Increased accuracy in locating the adsorbate atoms was achieved by tuning in not only to the fundamental (111) reflection of Ge, but also the higher order (333). This reduces the spatial periodicity of the XSW. At room temperature the Br atoms were found to vibrate in the [111] direction with a RMS amplitude of 0.067(53) Å, while the equivalent parameter in bulk Ge amounts to 0.084 Å.

A drawback of the generation of XSW by diffraction off nearly perfect crystals is that the periodicity of the SW is limited to the periodicity, d, of the diffracting plane, which rarely exceeds ~4Å. As a consequence, in experiments with adsorbates, this distance is the maximum vertical distance within which one can unambiguously locate a particular atom. If that distance surface-to-atom happens to be larger, one is left with a modulo-d ambiguity. To bypass this difficulty, one has recently resorted to the use of multilayers, where d can be more or less custom-designed for the experiment contemplated (~10Å<d<70Å).

An ultimate way-out is the substitution of total reflection for Bragg diffraction or the use of both. This has been recently shown by Bedzyk et al. [30], who reflected the SR beam off a gold-plated mirror. In contrast to previous experiments the substrate (=the mirror) was used to generate long period waves and only to support the compound of interest; the interaction between substrate and epilayer was of no particular interest. The object of study was the determination of the Zn ion concentration gradient in an electrolyte solution in contact with a phospholipid membrane - the resolution of a century-old problem in electrochemistry.

The grazing incidence surface diffraction technique was developped by Marra, Eisenberger and Cho [31] while studying the epitaxial growth of Al films on a GaAs substrate. Subsequent applications have focused on phase transitions in two-dimensional systems (in particular surface melting and surface ferroelectricity) and on the determination of surface layers. The success of the grazing incidence technique rests on the fact that for x-rays the refraction index of most materials is smaller than that of air, i.e. for angles below a certain critical angle the incident x-ray beam is totally reflected. Penetration into the bulk crystal is thus drastically reduced, and the signal one observes is due predominantly to surface diffraction. The scattering geometry is sketched in fig. 14.

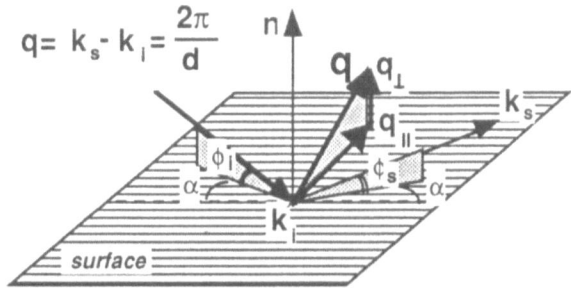

Fig.14 **Grazing incidence surface diffraction.** As seen from the surface the incident x-ray beam subtends an angle ϕ_i (~0.4 deg for a Cu

surface); the beam is diffracted through an angle α and leaves the surface at an angle ϕ_S. Note that due to the 2-dimensionality of the diffracting surface, the scattering plane needs to be almost coplanar with it. Most of the photon momentum , q, transferred in the scattering process is parallel to the surface (component $q_{||}$), yielding accurate atomic positions mainly as projections onto the surface. Integrated intensities are customarily measured by rocking the surface around its normal n. By contrast to scattering of SR from a bulk crystal, this surface normal lies in the horizontal plane. The methodology of the structure analysis of surfaces by x-ray diffraction does not differ much from the one developped for bulk crystals.

To illustrate the power of the technique, let us dwell on a specific example. The case in hand is the location of the chemisorption sites of oxygen on Cu (110); this knowledge is of fundamental importance in catalysis. First experiments with low energy ion scattering done a decade ago resulted in a model with oxygen atoms buried beneath an unreconstructed Cu surface. In the ensuing years this model was further refined on the basis of the same experimental evidence, yielding a reconstructed (2 x 1) surface (i.e. a doubling of the unit cell in the surface direction [10], and then eventually expanded into a model where every other [001] Cu row (in the fcc structure of Cu) is missing (See Fig. 15). By contrast, scattering with ions of higher energy suggested a model where the previously missing Cu atoms were present, but relaxed outwards from the surface, producing a corrugated Cu surface sheet. And, finally, some investigators argued whether the oxygen atoms bridging the [001] Cu atoms might not possibly reside above the first Cu layer, instead of below.

Much of the experimental evidence in favor of the various proposals was more of a supporting than of a decisive nature. It is only with the application of grazing incidence diffraction that this problem was clinched once and for all [32].

In the data collection strategy and structure solution methods developed for the crystallography of bulk crystals, the specimen is always assumed to be infinite in extent. This assumption is supported by the observation that 3-D diffraction scans of the reflected intensities show these to be spot-like. However, by restricting the irradiated area to the surface layer only, one observes that the scattering at positions of the reciprocal lattice of the bulk crystal becomes sharp only within the 2 dimensions of the surface, but diffuse in the direction perpendicular to it. Inclusion of the finiteness of the crystal in the theoretical prediction of diffraction demonstrates that these effects are due to the truncation of the crystal structure at the surface, hence the name crystal truncation rods (CTR). In principle, these diffuse scattering rods contain detailed atomic information on the termination of the surface.In cases where the surface atomic structure is completely unknown, one first searches for and measures both the integer-order reflections at bulk positions (CTR's) and any fractional-order reflections present. The latter's presence is indicative of surface reconstruction with a unit cell differing from the unit cell of an ideally terminated surface; analysis of the former yields models of the atomic structure of the layers immediately below the reconstructed surface layer. An F^2 Fourier synthesis (= Patterson or autocorrelation function) including only the intensities of the in-plane fractional-order reflections then reveals the projection of the atomic positions of the topmost, reconstructed layer onto the surface. This surface structure is necessarily related to the underlying bulk structure, but has only few atomic positions in common.

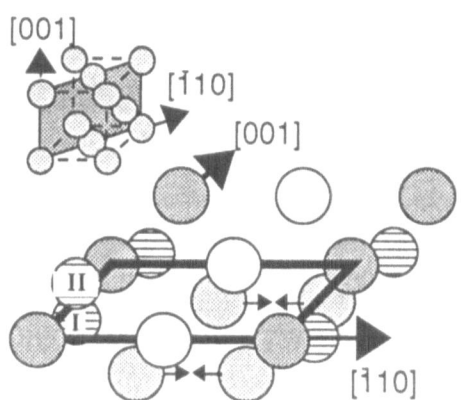

Fig. 15 Surface structure of oxygen chemisorbed on Cu (110) surface. The (2 x 1) surface superstructure ("recons-truction") is outlined. Cu atoms are shaded, where present, and left white, when absent; oxygen atoms are hatched (**I** denotes the now accepted site, **II** the discarded one). Arrows indicate the relaxation direction for Cu atoms in the 2nd layer. The inset (upper left) shows the Cu structure with the (110) plane shaded in.

In the example of interest, oxygen chemisorbed on Cu (110), the approximate surface structure was not in doubt; it had already been established on the basis of experimental evidence other than diffraction data. What remained to be done, was to unravel the details of the atomic rearrangement. The analysis of the integer-order reflections first confirmed the correctness of the missing row model for the Cu substructure. In addition, the F^2 Fourier synthesis including only the fractional-order reflections revealed 1) a lateral relaxation (by 0.031 Å) of the Cu atoms in the 2nd layer towards the missing Cu atom positions (see fig. 17) and 2) the presence of oxygen atoms at the (0,1/2) position of the reconstructed unit cell. Finally, a quantitative analysis of the CTR contributed the heretofore lacking information on out-of-plane displacement of the atoms in the last few layers. It turns out that the Cu atoms in the missing row layer is relaxed outward by 0.37 Å relative to the ideally terminated surface and that the oxygen atoms are located below the missing row layer at a depth of 0.34 Å.

For completeness's sake, it should be noted that the (2 x 1) surface reconstruction is not the only one observed on Cu (110); a (6 x 2) superstructure has recently been recognized [33].

If this had been a bulk crystal problem, one would have scoffed at the simplicity of the proposed structure when compared to the amount of experimental time invested. The difficulty in the solution of a surface structure resides less in the complexity of the analytical methods than in the faintness of the diffraction effects to be observed. Also, doing surface studies requires a heavy ultra-high vacuum chamber (preferably mounted on the diffractometer), the possibility of orienting the specimen within a second of arc, and the possibility of characterizing the surface layer with other, spectroscopic methods previous to the x-ray experiment.

While GI needs at minimum a surface of 25 mm^2, XSW fluorescence can do with 10^{-2} mm^2.

Most substances have an index of refraction less than unity for thermal neutrons, enabling total external reflection of neutrons at low angles of incidence. GI experiments with a neutron beam are conceivable; the main difficulty is the beam's low intensity.

SR *Due to the strong interaction between electrons and matter LEED is very surface sensitive and easily reveals this cell of the surface structure. It does not, however, yield the atomic positions within this cell - one of the key data needed in surface science. Indeed, the quantitative interpretation of the intensitites of the backscattered electrons is difficult; in contrast to x-ray diffraction, one is dealing with multiple scattering events (dynamical theory of diffraction) and with atomic form factors which are not well known.*

*On the other hand, although the interaction between x-rays and matter is much weaker than for electrons (and hence not particularly surface-sensitive), the use of grazing incidence obviates this disadvantage by probing only the top layers. The XSW technique, particularly in its total-reflection variant, and GI diffraction are both surface-sensitive (ppm of impurities) and readily interpretable. The GI technique, moreover, does not require the specimen to be nearly perfect. A condition sine qua non for the success of these techniques is the availability of **high intensity** (> high yield of fluorescence in XSW; higher counting rates in GI diffraction). This is essential -particularly for structures with light atoms- as the surface layers make out only a tiny fraction of the bulk crystal. The **small divergence** of SR facilitates grazing incidence experiments enormously; to obtain a similarly narrow collimation with a conventional source one would have to sacrifice a great deal of intensity.*

*The easy **tunability** of SR also contributes to the success of these techniques. In the case of XSW it permits one to always select a wavelength close to the absorption edge for the atoms of interest, and this increases the fluorescent* *yield. In the case of GI scattering, tuning to energies close to resonance increases the contrast for a particular atom (f").*

References

[1] J.R. Helliwell, Synchrotron X-radiation Protein Crystallography: Instrumentation, Methods and Applications, Rep. Prog. Phys. **47** (1984) 1403-1497

[2] D. M. Szebenyi, D. Bilderback, A. LeGrand, K. Moffat, W. Schildkamp and T.-Y. Teng, 120 Picosecond Laue Diffraction Using an Undulator X-ray Source, Trans. Amer. Cryst. Assoc. **24** (1988) 167-172 ; see also R. Pool, Molecular Photography with an X-ray Flash, Science **240** (1988) 295

[3] R. Bachmann, H. Kohler, H. Schulz, H.P. Weber, V. Kupcik, M. Wendschuh-Josties, A. Wolf and R. Wulf, Strukturanalyse an einem CaF_2-Einkristall mit nur 6 μm Kantenlaenge: Ein Experiment mit Synchrotronstrahlung, Angew. Chemie **95** (1983), 1013 [Angew. Chemie Int. Ed. Engl. **23** (1984)]

[4] R. Bachmann, H. Kohler, H. Schulz & H.P. Weber, Structure Investigation of a 6 μm CaF_2 Crystal with Synchrotron Radiation, Acta Cryst **A41** (1985) 35

[5] D.E. Cox, B.H. Toby and M.M. Eddy, Acquisition of Powder Diffraction Data with Synchrotron Radiation, Aust. J. Phys. **41** (1988) 117-131

[6] M. J. Hart and W. Parrish , Polycrystalline Diffraction and Synchrotron Radiation, Materials Res. Symp. Proc. **143** (1989) 185-195

[7] L.M. McCusker, The *Ab-Initio* Structure Determination of Sigma-2 (a New Clathrasil Phase) from Synchrotron Powder Diffraction Data, J. Appl. Cryst. **21** (1988) 305-310

[8] R.J. Cernik, P.K. Murray, P. Pattison and A.N. Fitch, A Two-Circle Powder Diffractometer for SR with a Closed Loop Encoder Feedback System, J. Appl. Cryst. **23** (1990) 292-296

[9] H. Iwasaki, T. Kikegawa, T. Fujimura, S. Endo, Y. Akahama, T. Akai, O. Shimomura, S. Yamaoka, T. Yagi, S. Akimoto and I. Shrotani, Synchrotron Radiation Study of Phase Transitions in Phosphorus at High Pressures and Temperatures, Physica B

[10] W.H. Keesom, J. de Smedt and H.H. Mooy, Proc. Kon. Akad.v.Wetens. Amsterdam **33** (1930) 814

[11] R.M. Hazen, H.K. Mao, L.W. Finger and R.J. Hemley, Single-crystal X-ray Diffraction of n-H_2 at High Pressure, Phys. Rev. **B36** (1987) 3944-47

[12] H.K. Mao, A.P. Jephcoat, R.J. Hemley, L.W. Finger, C.S. Zha, R.M. Hazen and D.E. Cox, Synchrotron X-ray Diffraction Measurements of Single-Crystal Hydrogen to 26.5 Gigapascals, Science **239** (1988) 1131-34

[13] N. Hamaya, Y. Fujii, S. Shimomura, Y. Kroiwa, S. Sasaki, and T. Matsushita, X-ray Diffraction Study of Phase Transitions in $[N(CH_3)_4]_2$ $MnCl_4$ under Pressure, Sol. St. Commun. **67** (1988) 329-332

[14] M. A. Subramanian, C.C. Torardi, J.C. Calabrese, J. Gopalakrishnan, K.J. Morrissey, T. R.Askew, R.B. Flippen, U. Chowdhry and A. W. Sleight, A New High Temperature Superconductor: $Bi_2Sr_{3-x}Ca_xCu_2O_{8+y}$, Science **239** (1988) 1015-1017

[15] P. Lee, Y. Gao, H.S. Sheu, V. Petricek, R. Restori, P. Coppens, A. Darovskikh, J. C. Phillips, A. W. Sleight and M. A. Subramanian, Anomalous Scattering Study of the Bi Distribution in the 2212 Superconductor: Implications for Cu Valency, Science **244** (1989) 62-64

[16] Y. Gao, P. Lee, P. Coppens, M. A. Subramanian, A. W. Sleight, The Incommensurate Modulation Study of the 2212 Bi-Sr-Ca-Cu-O Superconductor, Science **241** (1988) 954-956

[17] V. Petricek, Y.Gao, P. Lee and P. Coppens, X-ray Analysis of the Incommensurate Modulation in the 2:2:1:2 Bi-Sr-Ca-Cu Superconductor including the Oxygen Atoms, Phys. Rev. **B42** (1990) 387-392

[18] M. Spackman, H.P. Weber and B.M. Craven, Energies of Molecular Interactions from Bragg Diffraction Data, J. Amer. Chem. Soc. **110** (1988) 775

[19] Electron Density Mapping in Molecules and Crystals, ed. by F.L. Hirshfeld, Israel j. Chem. **16** (1977) 87-229

[20] The Application of Charge-Density Research to Chemistry and Drug Design, ed. by G.A. Jeffrey, NATO Advanced Study Institute Proc., North-Holland Publishing, (1991), in press

[21] H-R Höche, H. Schulz, H.P. Weber, A. Wolf, R. Wulf and A. Belzner, Measurements and Correction of Secondary Extinction in CaF_2 by means of Synchrotron Radiation X-ray Diffraction Data, Acta Cryst. **A42** (1986) 106-110

[22] F. S. Nielsen, P. Lee and P. Coppens, Crystallography at 0.3 Å: Single-Crystal Study of $Cr(NH_3)_6$ $Cr(CN)_6$ at the Cornell High Energy Synchrotron Source, Acta Cryst. **B42** (1986) 359-364

[23] A.Kirfel and K. Eichorn, Accurate Structure Analysis with Synchrotron Radiation. The Electron Density in Al_2O_3 and Cu_2O, Acta Cryst. **A46** (1990) 271-284

[24] B. Batterman, Effect of Dynamical Diffraction in X-ray Fluorescence Scattering, Phys. Rev. **133** (1964) A759-764

[25] P.L. Cowan, J. A. Golovchenko and M.F. Robbins, X-ray Standing Waves at Crystal Surfaces, Phys. Rev. Lett. **25** (1980) 1680-1683

[26] B. Batterman, Detection of Foreign Atom Sites by their X-ray Fluorescence Scattering, Phys. Rev. Lett. **22** (1969) 703-705

[27] K. Akimoto, T. Ishikawa, T. Takahashi and S. Kikuta, Structure Analysis of the $NiSi_2$/(111) Si Interface by the X-ray Standing Wave Method, Nucl. Instrum. Methods **A246** (1986) 755-759

[28] J. Zegenhagen, K-G Huang, B.D. Hunt and L.J. Schowalter, Interface Structure and Lattice Mismatch of Epitaxial $CoSi_2$ on Si(111), Appl. Phys. Lett. **51** (1987) 1176-1178

[29] M.J. Bedzyk and G. Materlik, Determination of the Position and Vibrational Amplitude of an Adsorbate by means of Multiple-Order X-ray Standing Wave Measurements, Phys. Rev. **B31** (1985) 4110-4112

[30] M. J. Bedzyk, G.M. Bommarito, M. Caffey and T.L. Penner, Diffuse-Double Layer at a Membrane-Aqueous Interface Measured with X-ray Standing waves, Science **248** (1990) 52-56

[31] W.C. Marra, P. Eisenberger and A.Y. Cho, X-ray Total-External-Reflection-Bragg Diffraction: A Structural Study of the GaAs-Al Interface, J. Appl. Phys. **50** (1979) 6927-6933

[32] R. Feidenhans'l, F. Grey, R.L. Johnson, S.G.J. Mochrie, J. Bohr and M. Nielsen, Oxygen Chemisorption on Cu (110): A Structural Determination by X-ray Diffraction, Phys. Rev **B41** (1990) 5420-5423

[33] R. Feidenhans'l, F. Grey, M. Nielsen, F. Besenbacher, F. Jensen, E. Laegsgaard, I. Stensgaard, K. W. Jacobsen, J.K. Nørskov, and R.L. Johnson, Oxygen Chemisorption on Cu (110): A Model for the c(6x2) Structure, Phys. Rev Lett. **65** (1990) 2027-2030

Outlook

SYNCHROTRON RADIATION: SOME POSSIBLE EXPERIMENTS WITH A THIRD GENERATION MACHINE

R. Rosei
Sincrotrone Trieste
Padriciano 99, I-34012 Trieste

The wavelength range and the magnitude of the radiation emitted by ELETTRA, the synchrotron radiation source being built near Trieste, will make it a unique instrument for research in physics, chemistry, biology, medecine and environment sciences.

Electromagnetic radiation, starting with light, the visible part of the spectrum, has always been an unreplaceable means of scientific investigation. Every time technology has managed to increase the quantity of available radiation for some analysis or experiment, an enrichment in our knowledge has followed. One could mention the invention of the telescope by Galileo, which permitted the discovery of the satellites of Jupiter, just increasing the magnitude of the observable light by a factor 10. More recently the invention of the LASER, which produces light beams of very high intensity and collimation, has been widely used for scientific and technological purposes which are also common to our daily life.

A few months ago in a landscape nearby Trieste (Carso triestino) a special machine for the production of beams of highest brilliance, particularly in the soft X-ray spectral region, has started being built.

When talking about soft X-rays, one means electromagnetic radiation whose wavelength is between 5 and 100 Å.

The magnitude and collimation of the beams produced by this machine will be so high that its features will resemble those of an X-ray Laser. These sources will be employed in order to make vanguard scientific experiments in many different disciplines and the possibilities offered for new and fascinating discoveries are so many, that the expectation for the construction of the machine is very big in the international community.

The machine which will produce these wonderful light beams is a third generation synchrotron radiation source called ELETTRA.

As a matter of fact, ELETTRA will not be a real synchrotron but a storage ring, made up by a steel doughnut having a perimeter of 260 m and the approximate shape of a dodecagon.

In this doughnut, bunches of electrons (previously accelerated by a LINAC) will be injected at a speed very close to that of light with an energy between 1,5 and 2 GeV.

As foreseen by the electromagnetic theory, electrons emit big quantities of sychrotron radiation tangentially to their orbit every time they are deflected by one of the bending magnets of the machine (which help keeping the electrons in their orbit inside the doughnut).

This process has been known for some time and it is usually considered as an inconvenience by elementary particle physicists. In fact they wish to accelerate electrons at increasingly higher energies (in order to use them as projectiles to generate subnuclear reactions), but the continuous loss of energy as synchrotron radiation makes the acceleration process harder and more expensive.

On the other side, synchrotron radiation extracted from lateral exits tangential to the doughnut of the accelerator, has attracted in the last 20 years an ever increasing number of experimentalists in different disciplines. These researches, after having employed for a long time the radiation produced by accelerating machines built for the investigation of elementary particles, have succeeded in obtaining the construction of some dedicated machines, that is, machines which are exclusively employed for scientific experiments using this light. In Berlin, Brookhaven, Paris and Tsukuba, big centres based on dedicated synchrotron light sources (known as second generation machines) have been built.

Science and technology give us the opportunity to make a big step further in the construction of not only dedicated machines, but also machines which are optimized in the production and exploitation of synchrotron radiation, having characteristics which will be uncomparably superior to those of the preceeding generation. ELETTRA will be one of the first machines of the new generation: a similar machine is in project in the famous Lawrence Berkeley Laboratories in California and another one in Grenoble, in France.

The electron beam which will turn in the doughnut (and which is the real source of the light) will have to follow some very stringent requirements: it will have to be very thin (few tens of a micron) and very collimated. These properties are not very easily obtainable and they keep some of the most qualified specialists of accelerating machines busy in the ELETTRA project.

The second ingredient, which is crucial for a high brilliance, is an apparatus called ondulator. Ondulators are made up by a very high number of alternated magnetic poles. These poles are placed in the 12 straight sections of the machine and they will give electrons a very narrow ondulating trajectory (from this the name of ondulators). This way, the radiation emitted by each edge of the ondulating trajectory will add and strengthen the radiation emitted by the next edge. The result of this coordinated action is impressive and will produce light beams whose peculiar characteristics of directionality, collimation and partial coherence will make it unique and very precious.

Other characteristics of not lower importance in ELETTRA will be a long lifetime and a great stability of the electron beam, so as to give sources of very stable light in order to allow for long lasting experiments under normal conditions.

Possible uses of these extraordinary sources for scientific experiments are very numerous and cover almost all experimental disciplines from physics to medecine, from chemistry to biology.

In the field of physics the most fascinating experiments will concern the most advanced materials such as layers of magnetic materials, manmade many layers semiconductors and the new high temperature superconductors.

It is interesting to mention in this context how physics and modern technology are getting to a target sought for centuries: that of managing to build, for each application, suitable materials. While in the past the prescription for making a harder clay pot or a more tenace steel blade came from empirical and casual observations, nowadays we are on the way of making projects for the construction of a material to make a transistor or an aeroplane turbine, establishing *a priori* its structure one atom after the other. ELETTRA can give a great contribution to this fantastic aim because its light beams, having wavelengths comparable to atomic and molecular dimensions, represent the most powerful means of investigation in order to check the growth processes of the new materials and verify the foreseen characteristics. Surely the most suitable technique for this type of studies is photoemission. It consists in sending a small monochromatic beam of light on the sample to be examined, and detect the electrons that the sample consequently emits.

The electrons that are emitted from a solid come from the chemical bonds that keep the material together and a measurement of their energetic distribution has all the necessary information so that the structure and the chemical and physical properties of the material under investigation can be understood.

It will be possible then to understand the details of the mechanisms which rule the formation of a new junction between a semiconductor and a metal, in order to get to the construction of electronic devices which are more compact, more reliable and having higher performances than those used nowadays. Furthermore one will manage to investigate the basic physical mechanisms of the new superconductors. Upon these basis one could possibly invent other materials which can be superconductors at room temperature too and which could also have other important characteristics such as malleability and ductility, so as to allow for practical applications unforeseeneable nowadays.

Another discipline which will gain many advantages from the experiments with ELETTRA light is chemistry. The technique which will be employed is E.S.C.A. (acronym for Electron Spectroscopy for Chemical Analysis). It is a sort of photo-emission in which one uses more energetic radiation which allows for the study of electronic states of the material having deeper energies.

The high brilliance of ELETTRA will give us the opportunity of focussing the radiation inside catalytic reactors of microscopic dimensions. We shall then be able to follow the course of a chemical reaction following the formation and transformation of all the intermediate species up to the final product. At the Trieste Synchrotron we have already received proposals for the study of the reaction between nitrogen oxide and carbon oxide (very important for cleaning the exhaust gases of cars), for the study of the reaction of sulphur oxide with calcium carbonte (important for the reduction of chimney smokes which are responsible for acid rains) and finally for the study of the reaction between chlorofluorocarbons and ozon (in order to better understand the mechanism which reduces this protective gas of our stratosphere).

The high brilliance characteristics of the electromagnetic radiation emitted by ELETTRA will make it relatively easy to focus this light until concentrating it on spots of a few hundred Å of diameter. This new possibility will open the way to a whole series of techniques of microscopy and X-rays which will have many advantages over conventional optical microscopy and over electron microscopy. For example an E.S.C.A. microscopy technique could have many applications in material sciences and, more in general, in applied and industrial research. With this technique one will be able to study the characteristics of strongly heterogeneous materials like catalysts, in which the active material is usually in the shape of microscopic granules deposited on a less reactive matrix. We shall then be able to discover the real chemico-physical composition of the catalyst in conditions similar to those of reaction and we shall then be able to make more efficient and selective catalysts with evident advantages for the chemical industry (which uses catalysts to produce the great majority of its goods, from plastic materials to fertilizers and synthetic textile fibers).

Furthermore one could study the morphology and the composition of metallic matrix materials like alloys and special cast irons, in order to understand the reasons why sometimes important and delicate components such as a turbin blade of the engine of an aeroplane or the boiler of a heating station can weaken of crack. Another microscopic technique which will surely have many interesting applications not only in industry but also in several other more fundamental disciplines is fluorescence spectroscopy. In this case, a strongly focussed X-ray beam will be sent on a sample which emits secondary X-ray radiation at different wavelenghts according to the chemical elements it consists of. One can then have a nondestructive chemical analysis of the composition of the sample even when it is strongly inhomogeneous. The most surprising aspect of this technique is its very high sensitivity so that one can detect also traces of elements below a millionth of a millionth of gram. The ecologic and environment physics applications of this technique are obvious since one will manage to reveal microscopic quantities of toxic elements like mercury, lead and cadmium in samples of water, trees, food or soil.

A more exotic, but also more amusing, application of fluorescence has been suggested for the study of archaeological remains. In many cases it is extremely interesting to make a non-destructive chemical analysis at high sensitivity of ancient pottery, because the distribution of the chemical elements contained in it is like a fingerprint which reveals the place of origin of the object. This way, knowing the composition of microelements in a brass object one can go back for comparison to the place where the foundry, which produced it, was. One can then obtain detailed information on the ancient commercial roads. This last example shows one of the most surprising consequences of synchrotron radiation which are its applicability to the most different and sometimes unthinkable disciplines.

Many of the examples of using ELETTRA for scientific experiments are obtained by the extrapolation of what we can already obtain today with our present machines, taking into account that its brilliance will at least be one thousand times higher. But, so as it was for the galilean telescope, which lead to an unforseen discovery just increasing light by a factor then, we are sure now that the most fascinating applications of this wonderful machine are far from being foreseeable.

Index